D1083315

Supernormal Stimuli

Also by Deirdre Barrett:

Waistland

Trauma and Dreams

The Pregnant Man and Other Cases from a Hypnotherapist's Couch

The Committee of Sleep

The New Science of Dreaming

Supernormal Stimuli

How Primal Urges Overran Their Evolutionary Purpose

Deirdre Barrett

W. W. Norton & Company
New York • London

Printed in the United States of America
First Edition

For information about permission to reproduce
selections from this book, write to Permissions,
W. W. Norton & Company, Inc.,
500 Fifth Avenue, New York, NY 10110

For information about special discounts for bulk
purchases, please contact W. W. Norton Special Sales at
specialsales@wwnorton.com or 800-233-4830

Manufacturing by The Courier Companies, Inc.
Book design by Judith Stagnitto Abbate / Abbate Design
Production manager: Andrew Marasia

Library of Congress Cataloging-in-Publication Data

Barrett, Deirdre.
Supernormal stimuli : how primal urges overran their
evolutionary purpose / Deirdre Barrett. — 1st ed.
p. cm.
Includes bibliographical references and index.
ISBN 978-0-393-06848-1 (hbk.)
1. Evolutionary psychology. 2. Behavior evolution. I. Title.
BF698.95.B36 2010
155.7—dc22

2009037078

W. W. Norton & Company, Inc.
500 Fifth Avenue, New York, N.Y. 10110
www.wwnorton.com

W. W. Norton & Company Ltd.
Castle House, 75/76 Wells Street, London W1T 3QT

1 2 3 4 5 6 7 8 9 0

Contents

Chapter 1 • What Are Supernormal Stimuli? 1

Chapter 2 • Making the Ordinary Seem Strange 6

Chapter 3 • Sex for Dummies 29

Chapter 4 • Too Cute 52

Chapter 5 • Foraging in Food Courts 75

Chapter 6 • Defending Home, Hearth, and Hedge Fund 105

Chapter 7 • Vicarious Social Settings from Shakespeare to *Survivor* 132

Chapter 8 • Intellectual Pursuits as Supernormal Stimuli 159

Conclusion: Get Off the Plaster Egg 176

Acknowledgments 179

Notes 181

Illustration Credits 199

Index 201

Supernormal Stimuli

1

What Are Supernormal Stimuli?

The European Cuckoo, whose distinctive call issues from our cuckoo clocks, seems more goofy than sinister. However, this creature is the leading example of a brood parasite. A female cuckoo will sneak into the nest of another species when the parent bird is away and lay an egg, shoving a rightful one out so the count will be correct. She flies away to repeat this in other nests, leaving the care of her progeny to the unsuspecting adoptive parents.

The cuckoo egg resembles those of the host but it's often a bit larger or brighter. The nest's owner sits on the cuckoo eggs preferentially if there are too many to keep warm. When the baby cuckoo hatches, its beak is wider and redder than the other chicks'. *If there are any other chicks.* The mother cuckoo has already dumped one egg, and the baby cuckoo tries to push remaining eggs or newly hatched siblings to their death. The

Baby cuckoo fed by foster parent.

cuckoo enacts the Cinderella story in reverse: the stepmother gives Cinderella all the attention while her ugly stepsisters go wanting.[1]

Tragicomic dramas play out in every arena of animal life. Put a mirror on the side of a beta fighting fish's aquarium and the gaudy iridescent male will beat himself against the glass attacking a perceived intruder. A hen lays eggs day after day as a farmer removes them for human breakfasts—30,000 in a lifetime. Not a single chick hatches but she never gives up trying. Male barn swallows have light brown chests and females choose the ones with the most intense color as an indication of fitness. Scientists with a $5.99 felt-tip marker can darken the chest of a previously scorned male, and suddenly females line up to mate with him.[2]

These animal behaviors look funny to us . . . or sad . . . the reflexive instincts of dumb animals. But then there's a jolt of recognition: just how different are our endless wars, our modern health woes, our romantic and sexual posturing?

Human instincts were designed for hunting and gathering on the savannahs of Africa 10,000 years ago. Our present world is incompatible with these instincts because of radical increases in population densities, technological inventions, and pollution. Evolution's inability to keep pace with such rapid change plays a role in most modern problems.

Animal biology developed a concept that is crucial to understanding the problems instincts create when disconnected from their natural environment—that of the *supernormal stimulus*. Nobel laureate Niko Tinbergen coined this term after his animal research revealed that experimenters could create phony targets that appealed to instincts more than the original objects for which they'd evolved. He studied birds that lay small, pale blue eggs speckled with gray and found they preferred to sit on giant, bright blue ones with black polka dots. The essence of the supernormal stimulus is that the exaggerated imitation can exert a stronger pull than the real thing.

Supernormal stimuli explain why other birds give the baby cuckoo more attention than their own young; why the unyielding, invulnerable fish in the mirror provokes such a violent attack; why the faux-feathered lothario gets all the girls.

"You have a club—use it."

Many evolutionary concepts have been applied to human behavior by biologists. Some have crossed over into popular conversation. However, the importance of supernormal stimuli has not yet been fully appreciated in either arena—until now. In the pages that follow, I appropriate the term to explain a broad array of human folly.

Animals encounter supernormal stimuli mostly when experimenters build them. We humans can produce our own: candy sweeter than any fruit, stuffed animals with eyes wider than any baby, pornography, propaganda about menacing enemies. Instincts arose to call attention to rare necessities; now we let them dictate the manufacture of useless attention-grabbers. My last book, *Waistland*, explored how supernormal food stimuli have produced our obesity crisis. The present book applies the concept to sex, health, international relations, and media. I also draw on other ideas from animal ethology and evolutionary psychology that further illuminate the disconnection of instincts from their natural environment.

It's not all bad news. Once we recognize how supernormal stimuli operate, we can craft new approaches to modern predicaments. Humans have one stupendous advantage over other animals—a giant brain capable of overriding simpler instincts when they lead us astray. This book emphasizes the importance of recognizing when we need to push the override button.

Chapters 3 through 8 each examine a particular modern problem. But first, Chapter 2 describes how Tinbergen discovered supernormal stimuli and related concepts. I devote significant space to this man's quirky life story because his distinctive personality, his extraordinary family, and the events of World War II all contributed to why he was able to step outside the perspective of his culture—and at times his species—to examine human instincts and behaviors with such remarkable objectivity. Chapter 2 concludes with a summary of how ethology, evolutionary anthropology, and psychology have built on Tinbergen's insights in the years since.

2

Making the Ordinary Seem Strange

Born in 1907 in The Hague, Niko Tinbergen grew up as the second of five in a family that would produce what none other has before or since: two Nobel Prize–winning siblings. Tinbergen's parents were both schoolteachers. They showered praise on studious oldest brother Jan, while Niko considered his youngest brother, Luuk, the brightest in the family. As a middle child, Niko "only just scraped through, with as little effort as I judged possible without failing."[1]

The main biography of Tinbergen, written by his former graduate student, Hans Krunk, is strongest in describing how his scientific theories developed.[2] Krunk takes Tinbergen at his word in portraying the family as "happy and harmonious," his father as "a hard worker and intellectually stimulating," and his mother as "warm and impulsive." There is frustratingly little information to explain how the family developed rifts so profound that some didn't speak to others for decades or why they ignored major depressions among their members and even the eventual suicide of one.

More is known about Tinbergen's early passion for watching fish and fowl. In a pond in the Tinbergen backyard, Niko could observe stickleback fish, frogs, and waterbirds. He roamed nearby woods and beaches, sometimes with little brother Luuk in tow, finding more exotic wildlife.

Niko joined the *Nederlandse Jeugdbod voor Natuurstudie* (NJN; roughly translated as the Dutch Youth for Nature Studies).This club organized hikes and summer camps—like the Boy Scouts, except members were of both genders aged 12–23, and there was no adult supervision. "Boys and girls mixed freely—highly unusual, even as late as the 1950's, but anything sexual was frowned upon," a fellow NJNer recalls. "We were very chaste, and established couples were not expected to even hold hands in company."[3]

Dutch society of this era opposed hunting; wildlife had been decimated by early settlers and was only just recovering. Niko embraced this policy in his youth. He stalked animals only to photograph, draw, or simply observe them, but he would later come to describe this as an offshoot of the natural instinct for hunting.

Niko's images of birds appeared in NJN's annual calendars and he wrote for their magazine, *The Amoeba*. One early article anticipated his focus on the relevance of animal behavior for humans: "With careful observation of the intimate life of the birds, you recognize yourself in all their expressions, and that is the value of birding. However strange it may sound, everyone best recognizes his own faults when he sees them made by someone else."[4] Luuk followed Niko's lead, joining the NJN and beginning to draw for *The Amoeba*.

Upon graduation from high school, oldest brother Jan enrolled at the University of Leiden, the best school in the Netherlands. He commuted from home as he studied physics. Niko was initially unsure whether he wanted to attend college. He considered becoming a professional photographer or forest ranger. These nonacademic possibilities alarmed his parents, and they persuaded a biologist friend to employ him at a birding reserve after graduation. Two months there convinced Niko that he could pursue his fascination with wild animals while earning a degree in biology, so he enrolled at Leiden halfway through the fall semester. He remained a lackluster student but continued in the biology doctoral program.

Toward the end of graduate school, Niko discovered one of the largest known colonies of bee-wolves, a type of digger wasp. Named for their prey, they intrigued him with their skillful hunting of other stinging insects. Niko wondered how the wasps could fly so far afield and yet unerringly locate their tiny burrow upon returning. Darwin's writings of fifty years before were well accepted by Dutch biologists, but most had focused on his ideas about the evolution of structures. The conventional approach to assessing what animals could see or smell was to dissect nostrils and eyes. Tinbergen focused on Darwin's assertion that behavior had evolved intricacies equal to those of structure and much could be learned by observing animals in action.

Niko placed pinecones around the wasp nests, leaving them for a few hours of training. Then, while the wasps were out, he moved the cones to a new mound. A wasp would return with a bee, land amid the cones, and search in vain for the hole. Niko found that placing inconspicuous pine-soaked cardboard by the entrance and moving that did not result in the wasps following it to a new location, proving smell was not a factor. Brightly colored paper moved around did not influence them either—their cues were visual but they only noticed three-dimensional objects and not color.

This systematic method of intervening one stimulus at a time, without otherwise disrupting the lifestyle of wild species, would revolutionize what could be learned by biologists. When it came time for Niko to write a dissertation, however, he summed up the bee-wolf results simply, underplaying the novelty of his approach. His dissertation was 29 pages long. "It must have been one of the shortest theses in the history of biology," observes biographer Krunk. "Several hundred pages then was the norm."[5] Tinbergen was barely awarded his PhD after much debate among his professors, but he gained something academic honors could not have given him. Krunk wrote, "If Niko had been taken on for a PhD in some strong research group, he would not have developed the idea for field experiment the way he did. He followed his own lead, set his own questions, and used his own arguments. The Leiden professors tolerated, but hardly guided him."[6]

Upon graduating, Tinbergen obtained a fellowship to take part in an expedition to study wildlife in Greenland. He married a young woman he'd met in the NJN, Lies Rutten, who followed him northward. Niko and Lies soon left the rest of the scientists and traveled up a small fiord to an area inhabited by Inuit, then referred to by westerners as "Eskimos," and lived there for the year.

Most Dutch would have been lost on the dazzling blue-white tundra. Hunting and fishing were the main source of food, dogsleds and kayaks the only transportation. But Niko and Lies thrived. The Inuit taught Tinbergen to paddle a kayak and right an overturned one in icy water. He became a skilled marksman, using a traditional Inuit white screen to stalk seals, returning upstream with the carcasses on the back of his kayak. When the Inuit killed many more birds or seals than they bothered to retrieve, Tinbergen recorded this matter-of-factly, never criticizing their practices even in his private diary. Despite his interest in animals, he wasn't inclined to make friends, much less pets, of them. The Inuit attitude toward sled dogs as captive work animals, beating them and killing them once they'd outlived their usefulness, never disturbed him.

Some of Tinbergen's habits appeared eccentric to the Inuit, however. He spent hours recording the movements of birds—which were no more special to the Inuit than rocks or blades of grass. The Inuit view of animals as intricate but soulless machines may have influenced Tinbergen's later science—it certainly dovetailed with the approach he was developing. His budding field of ethology simply described behavior, never speculating on internal states.

Tinbergen's grant was to study snow buntings. Other closely related buntings were known to be highly territorial, but a previous biologist visiting Greenland had reported that the snow bunting did not defend territory. It soon became apparent that this predecessor was wrong: he'd been there only part of one year. Tinbergen discovered that territorial behavior was tied to reproductive season: it mattered to the males when courting mates and to both genders when feeding their young. This association of territoriality to resources was a theme to which Tinbergen would return with other species. We revisit it later in this book in light of how humans—originally nomadic—have become so territorial.

Tinbergen also observed ground-nesting terns and became interested in the fact that, if an egg was dislodged from the nest, the terns rolled it back into place. This seemed an opportunity to test how terns recognized their own eggs much as he'd experimented with how wasps recognize their burrow. Tinbergen placed eggs of other birds near the terns' nests, finding they would retrieve a variety of colors and sizes. This model eventually led to the discovery of supernormal stimuli, but Tinbergen played with it only briefly that year, realizing this research could be continued with gulls back home.

When Tinbergen returned to Holland, his brothers were already excelling in academic pursuits. Jan had written a dissertation (a lengthy one) on the application of certain principles from physics to economics and was established as an economics professor in Amsterdam.[7] Luuk, as a 19-year-old undergraduate in biology at Leiden, published a book on identifying birds, illustrated with his own drawings.

Niko became an instructor in Leiden's biology department and at this point he began to come into his own. He published his observations on the birds of Greenland for both scientific journals and popular magazines. Laden with Inuit artifacts and tales of life on the tundra, he and Lies were exotic and sought-after among the academics of Leiden. Tinbergen developed a course for undergraduates on animal behavior and led a field practicum for graduate students, who rushed to work with him. He set one group to observing the colony of wasps from his dissertation, another to continuing egg-rolling experiments with geese.

Tinbergen and his students also studied the stickleback—the most common freshwater fish in Holland. The male has a brightly colored red underbelly; it vigorously defends a territory and builds a nest in the center. Other males who enter the territory are attacked, but females are guided toward the nest to lay eggs. Sticklebacks continue these rituals when scooped out of ponds and installed in aquariums.

Tinbergen and his students constructed model fish and spent hours leaning over the tanks, "swimming" the dummies around. At first these were just dead fish on wires, but later they carved and painted wooden ones. Systematically varying the wooden models, they established that it was redness of the underbelly that signaled "male to attack." Sticklebacks didn't attack a realistically shaped model if its belly wasn't red, but violently pursued *very* unfishlike shapes with red undersides. Males in aquariums by the window went into attack mode when a red *postal van* drove by.

Color wasn't the determinant for detecting females; males escorted carved wooden models to the nest if they had the curved belly of an egg-bearing female. They preferred the model with the roundest stomach.

The most interesting of Tinbergen's discoveries was that dummies could surpass the power of any natural stimuli. Male sticklebacks ignored a real male to fight a dummy brighter red than any natural fish. They'd choose to escort an exaggeratedly round-bellied model over a real egg-bearing female. Simultaneously, Tinbergen and other students studied geese and found similar patterns. The characteristic that determined which egg a goose would roll back into the nest—color, size, markings—could be exaggerated in dummy eggs. The graylag goose ignored its own egg while making a heroic attempt to retrieve a *volleyball.*

Niko Tinbergen painting dummy eggs.

Luuk joined Niko in experimenting with a variety of birds. Song birds abandoned their pale blue eggs dappled with gray to hop on black polka-dot Day-Glo blue dummies so large that the birds constantly slid off and had to climb back on. Once a chick hatched, parents preferred to feed a fake baby bird beak on a stick if the dummy beak was wider and redder than the real chick's. Hatchlings begged a fake beak for food if it had more dramatic markings than their parents'.

Tinbergen named the phenomena of his exaggerated dummies "supernormal stimuli." He published papers outlining the concept that instincts were coded for a few traits and amplification of these traits could easily fool animals. While he was working on supernormal stimuli, Tinbergen read a theoretical article, "On Instinct," by a young German biologist, Konrad Lorenz. Intrigued by this broader framework for his experiments, Tinbergen persuaded the department to invite Lorenz to lecture.

Lorenz was "a large man in every sense," according to biographer and former student Krunk, "with a shock of hair and beard—an ebullient extrovert, always the soul of the party who needed to impress."[8] His style was as flamboyant as Tinbergen's was understated. Lorenz was "unashamedly anecdotal, living proof that eccentric inspired guesses are frequently the basis of scientific progress," recalls another former student, Desmond Morris. "His whole life-style seemed to be an animal-infested chaos."[9] Unlike Tinbergen, Lorenz kept pets—and not just standard ones like dogs. Goats and geese overran his house. Different as the two men were, they shared an endless fascination with animal behavior and bonded immediately.

Lorenz invited Tinbergen and his family to visit him outside Vienna. They stayed four months. "Whenever they were together in the same room, the air would be full of their tales and roars of laughter," recalled Krunk. They studied geese and discovered what they termed "an instinct-releaser": when they rolled an egg out of a goose nest, the mother began motions to retrieve it. These continued even if someone took the egg away while she was halfway through the process. They discovered "imprinting"—a few hours early in the life of goslings when whatever is in front of them becomes what they will follow throughout their youth. Usually this is the mother goose, but broods imprinted on Lorenz's wading boots and followed him in single file just as other broods did their mothers.

Ethology texts to this day feature photos and sketches of Lorenz with the ducklings trailing him, wild birds perching on him, small mammals draped around him as he clowns and experiments. These exist only because, for those four months, the flamboyant Lorenz was constantly in the company of the quiet photographer and illustrator Tinbergen—and Tinbergen was generating half the ideas for the antics. "This summer with Niko Tinbergen was the most beautiful of my life,"

Konrad Lorenz with his imprinted geese in a photo (*left*) and in his whimsical self-portrait (*right*).

Lorenz later wrote. "What we did scientifically had the character of play and, as Fredrich Schiller says, 'Man is only then complexly human when he is at play.' Niko and I were the perfect team."[10]

Back in Leiden, Tinbergen continued these explorations, now grounded in Lorenz's theoretical concepts. He studied butterflies and found that marks on the torso of the female and its vibratory movements were the sole mating releasers—he could construct dummy butterfly torsos with brighter stripes and more regular movements. Males would ignore a live, receptive female to mount cardboard cylinders that didn't even need wings!

The friends' careers and the new field of ethology seemed launched on a glorious trajectory. But it was not to be that simple. While Tinbergen and Lorenz had spent the summer of 1937 outside Vienna observing territoriality and aggression in geese, Adolf Hitler was building up Germany's military and eyeing the rest of Europe.

The following year, the Nazis gained influence in Austria and Lorenz was pressured to join the party to keep his job; he did. With the "Anschluss" German annexation of Austria, the

Catholic influence on the government crumbled and barriers to Darwinism disappeared with it. Lorenz found it easier to get funding under the new regime. He published papers on the effects of domestication on animals and men with some overtones that probably appealed to Nazi views on race though there was nothing patently offensive or scientifically wrong in the papers. He hoped this was as far as the Nazi juggernaut would impact him. But a year later, he was drafted. On a form asking what special skills he had, he listed motorcycle mechanics: that would have kept him away from the battlefield. Instead, the Nazis decided he'd studied enough anatomy and psychology to be a medic and sent him to the front lines.

Meanwhile in Holland, only the eldest son of each family was subject to the draft. Jan declared himself a conscientious objector. This was not complacency about the Nazis nor was it cowardice. (Lorenz's behavior may have been either or both.) Jan advocated the economic principle that violent confrontation was always destructive for both sides, that there were always two losers. He proposed radical economic sanctions as the only way to stop Hitler; at this point, the U.S. and many other countries were still trading with Germany. Niko, immune from the draft once Jan's C.O. status was approved, continued teaching until the Nazis invaded Holland. Leiden University was first to clash with them over attempts to fire all Jewish faculty. Niko was at the front of the protests. The Nazis closed the university; Niko and twelve other faculty members were sent to Beekvliet, which history has referred to as a "hostage camp" rather than a prison camp.

Beekvliet was a surreal place: a converted Catholic seminary. It was elegant in its common areas but men slept in modest dormitory-style housing, six to a room. Before Niko arrived, three men had been taken from their beds and shot in retaliation for railway sabotage by the Dutch resistance. One more prison inmate was executed after Niko arrived but then the killings stopped. The Nazis were holding powerful Dutch there as hostages to intimidate the country, but they were treated better than most "free" Dutch. After those first killings, the threat of execution faded into the background. Hot meals were served—continuing even into Holland's "winter of starvation," when the rest of the country was lucky to find stale bread. There was hot running water for daily baths.

Hostages were allowed to engage in whatever pursuits they liked. The camp owned a Steinway piano and several world-class musicians imprisoned there gave regular concerts. A famous hostage artist offered a class in portrait drawing which Niko attended. Inmates organized lectures on electromagnetism, Dutch history, Sophocles, and hieroglyphics. Niko lectured on his findings in animal behavior. "If you did not know that you are a hostage, you would think yourself at some interesting conference,"[11] Niko wrote in one letter home. He completed two books there: one a children's version of the stickleback story, illustrated with his drawings and written largely for his own children—though it was eventually published. The other was an academic work on instinctual releasers and supernormal stimuli. He met a publisher imprisoned at Beekvliet who accepted it, so publication was underway by the time he was released.

When Lorenz heard that Niko was in a hostage camp he wrote to Lies, offering to intercede with the Nazis. She wrote back icily that they wanted no Nazi help. Lorenz was not in a position to help anyone for long. He was captured by the Russians.

Unlike Beekvliet, Lorenz's camp did not allow letters in or out, so most of his family and colleagues, including Tinbergen, thought him dead for three years. But Lorenz survived tolerably. "I was never in a really bad camp," he later wrote. "If the prisoner kapo was an honest man and the Russian officer too you could live in perfect health."[12] Lorenz was well-liked by fellow prisoners and guards—though he intimidated a few. One exploit in particular added to his reputation. Seeing him about to catch a large tarantula, a Russian guard warned him of the danger, saying that it was very poisonous. Lorenz promptly picked up the spider and, safely gripping the head and thorax, bit off and consumed the fleshy abdomen. This sent the guard "running and screaming into the steppes of Kazakhstan," according to Lorenz. Spider eating was not all display; Lorenz believed the extra protein was a good addition to the prison diet. He was already in the habit of tasting whatever bugs were eaten by the birds he studied. He derided western scientific expeditions which sometimes starved after losing their supplies—there was actually food all around them.

There was no paper at Lorenz's camp—that was one of the reasons there were no letters out. But he was as determined as Tinbergen to continue his work and he wrote on emptied dark flour sacks with a pen filled with bleach. There was not much wildlife to observe, but he applied his scientific method to whatever appeared. All of the prisoners were infested with fleas. Lorenz noticed that some of his fleas occasionally began pirouetting, unlike any of their other movements. Consistent observation established that it was male fleas that did this and that a female had always just landed on that area of his skin. It was a flea courting dance. He wrote a paper about this that was later published. (It's a phenomenon that "trained" flea circuses put to good use.) But mainly Lorenz wrote a book reflecting on the research he and Tinbergen had done together. When the war finally ended and German prisoners were released, Allied gatekeepers along the road home noticed his contraband sack-manuscript. One made him sign an oath swearing they were indeed scientific papers, but he persuaded all to allow the German document to pass.

Tinbergen was released from Beekvliet shortly before the end of war. He joined Lies and the children in the countryside where they were living with her sister's family. Before the invasion, Luuk had married a young Jewish woman. The couple was in hiding, eventually occupying Niko's house in the city. Luuk and his wife were in grave danger, and many people Tinbergen knew died in the war. One of his old professors perished in the famine and the graduate student who spearheaded the stickleback research was killed in a battle with the Germans. However, everyone in the Tinbergen family survived until the Nazis retreated.

Leiden University reopened and science resumed. Tinbergen heard that Lorenz was alive, but the two made no attempt to contact each other, even as books based on their joint research came out and they continued their related careers. In 1945, Tinbergen wrote to a Dutch colleague with uncharacteristic emotion:

> It is impossible for me to resume contact with him [Lorenz] or his fellow-countrymen, I mean it is psycho-

> logically impossible. The wounds of our soul must heal, and that will take time. . . .This is not a result of a desire for revenge, but we simply cannot bear to see them.[13]

Lorenz wanted out of Austria-Germany. It was occupied by Allied forces, and devastated from years of bombing. The two countries were dividing again—and Germany was dividing further into East and West. Universities were in chaos. The multilingual Lorenz applied to jobs around Europe and the U.S. However, Tinbergen wasn't alone in his sentiments about Nazi academics. They were boycotted everywhere—except for rocket scientists or nuclear physicists who were snatched up, no questions asked. The upbeat Lorenz was unfazed and worked full-force as few German scientists did through this time. He published *King Solomon's Ring*,[14] whose title alluded to the legendary ring that gave King Solomon the power to understand animals. Lorenz claimed, metaphorically, that this was what ethological research achieved. Some of his opinions—such as his enthusiasm for pets—would have dismayed Tinbergen, but mostly the book stayed close to their mutual work and conclusions.

Niko and Luuk began to suffer depressions—periods of dark despair and suicidal thoughts that lasted at first for days, later for months at a time. But their careers thrived. Luuk became known for applying statistics to ethology. Meanwhile, Jan's highly statistical economics, called *econometrics*, rocketed him to international prominence; he consulted to the League of Nations. Luuk conferred with Jan and adapted some of his concepts to biology. Biographer Krunk hypothesized that Niko's blatant disregard for mathematical methods resulted from sibling rivalry with Jan and saw it as the only shortcoming in his ambitious expansion of ethology.

Tinbergen explored every other aspect of the field. He set up elaborate wildlife observation sites, modeled on his beloved NJN camps, for students. When Tinbergen was present, the experiments with animals occupied every waking hour and the students acted as if they were in a remote wilderness. When he left, the students enjoyed nightly parties and trips to bars near the camps. They chafed at Tinbergen's restrictions but nevertheless enjoyed his presence because of his devoted mentoring. "More than one of Niko's children said to me later that Niko was more interested in his students than in his family," reports Krunk.[15] "He found it difficult to do father-like things with us children," recalled one child. Said another, "He was so interested in his work that he regarded us as a nuisance." Even when Niko was at home, his presence focused on ethology. "I certainly had the feeling that [father was] observing me as an animal going through the courting-mating phase, and it wasn't a very comfortable feeling to have," said one of his daughters.[16] Luuk was the one family member with whom Niko was close—because of their shared research and perhaps also because of the shared depressions.

A year after the war ended, Niko was promoted to full professor; he was only 39. His award lecture for the occasion, "Nature Is Stronger than Nurture," warned against religion: "The emphasis by Christianity on our responsibility for our behavior has had the consequence that the differences between man and animal are perceived as too prominent."[17] Tinbergen had never been a religious man. Wartime atrocities, however, had highlighted the absence of a deity for him while both sides invoked one aligned with themselves, and this turned him into a militant atheist. When nine-year-old-daughter Toos wanted to go to church with a friend, the answer was an emphatic "No!" Krunk recalls Tinbergen being furious when he insisted on taking off an hour from field observation on Christmas Day to attend mass.

Niko began to find Holland too small academically. He was more successful than Lorenz in his job search, fielding offers from around the U.S. and Europe before settling on Oxford. Luuk visited him in England for months at a time, but Niko didn't bother to stay in touch with the rest of the family after his parents passed away. At Oxford, he taught dozens of the next generation of evolutionary biologists including two who became household names because of their popular writing: Desmond Morris who wrote *The Naked Ape*[18] and Richard Dawkins, noted for biology books beginning with *The Selfish Gene*[19] and for militant atheism similar to his mentor's. Tinbergen established new wildlife camps—some in England, others as far afield as Africa.

Shortly after the move to Oxford, Tinbergen attended a conference that Lorenz also attended. They saw each other across the room at a party the first night and immediately reunited. Tinbergen announced publicly to Lorenz, "We've won!"—meaning their friendship and work relationship had survived the war. A lively correspondence resumed and Tinbergen's support helped influence other biologists to disregard Lorenz's Nazi past.

In 1955, Luuk had achieved international prominence in ethology but his depressions worsened. At age 39, he took his own life. Niko didn't write of this in his journals or speak to anyone who passed along his reaction to biographers. Lorenz's response is recorded. He heard that the famous ethologist Professor Tinbergen had died by his own hand and was stricken, thinking this meant Niko. Hours later, he was relieved to learn it was Luuk, whom he knew only in passing. Later, when Luuk's son Joost came to Oxford for a Master's degree in biology under Niko, they barely mentioned Luuk. "The Tinbergen's are just like that," observed Joost, "very bottled up."[20]

Niko's own depressions worsened. "I was so far in Nairobi," he wrote after one field trip, "that four times I awoke at night and found myself with enough poison in my hand and a glass of water, to finish myself off completely."[21]

In 1969, at the age of 66, Jan Tinbergen received the Nobel Prize in Economics. Niko didn't break his silence to send congratulations. Four years later, at exactly the same age—66—Niko was awarded the Nobel Prize in Biology with Konrad Lorenz and another academic. Reporters swarmed around the Tinbergen brothers, wanting interviews with the laureate siblings. Photographers asked how they could get a photo of them together. Niko predicted this would never happen as neither was going to travel to the other. Jan sent a warm letter of congratulations, however. He'd been reading Niko's books and praised them. Niko never bothered to look at Jan's work but he did write back.

The photographers actually got their photo when Niko was in Amsterdam to receive the leading Dutch science prize en route to Geneva for the Nobel. Jan attended and the two chatted and posed together. The Nobel ceremony was more complicated. Some scientists balked at including Lorenz in the prize because of his wartime involvement. Tinbergen suggested to his friend that he apologize for his Nazi affiliation in his Nobel speech. Lorenz reportedly agreed, but he never made such an apology in any context. Tinbergen chose to organize his own speech around his recent interest in the disorder of autism. This was received poorly, lying quite outside his expertise. It's interesting he focused on autism in later years. Tinbergen obviously suffered from depression, but descriptions of him also have hints of Asperger's syndrome, the mildest form of autism, in which a person's capacity for empathy and socialization is diminished. It's probably no coincidence that such a cool personality was able to discover mechanistic principles governing instinctive behavior. Someone empathizing more with birds caring about their young might not have thought to check out whether parent birds would feed a fake beak.

Most of the scientific community disregarded both the Nazi history and the autism speech and focused on the content of the award. A *Science* editorial observed that the prize decision "might be taken . . . as an appreciation of the need to review the picture that we often seem to have of human behavior as something quite outside nature, hardly subject to the principles that mold the biology, adaptability, and survival of other organisms."[22]

Biologists on Mankind

Addressing the question, "What is Man?," biologist Richard Dawkins wrote: "There is such a thing as being just plain wrong and that is what before 1859 all answers were,"[23] referring, of course, to Darwin's *On the Origin of Species*. Darwin wrote about humans mostly in terms of their expression of emotion and its commonalities with the rest of the animal kingdom's.[24] Tinbergen clearly thought ethological principles applied to man but devoted the occasional sentence or two to him while waxing eloquently for pages about wasps and sticklebacks. Lorenz wrote more specifically of human dilemmas in his last books. *On Aggression* expanded his ideas about territoriality and dominance hierarchies of animals to explain human wars. *The Eight Deadly Sins of Civilized Man*[25] described how man's ability to manipulate his environment in ways no other animal can wreaked havoc.

In 1975, Harvard biologist E. O. Wilson's landmark *Sociobiology: The New Synthesis*,[26] used that term rather than "ethology" to describe how social behavior arose from evolution. He incorporated Tinbergen and Lorenz's work and added new observations from his own studies of ants. In *On Human Nature*, Wilson explored more explicitly what ants and other insects had to tell us about ourselves. Ants may seem strange kin with their simple brains and nervous systems but they live in highly complex social groups. The sterility of certain classes of workers and "armies" willing to sacrifice themselves to protect the nest led Wilson to theories of altruism applicable to human behavior. He was one of the first to suggest that homosexuality in certain circumstances conferred survival value under an "uncle effect."

Sociobiology as it applied to animals was immediately well-accepted; it was the inclusion of human beings that was controversial. The greatest furor was over Wilson's assertions that some of the individual variation on traits such as intelligence, creativity, and introversion versus extraversion is genetic. He believed men and women had biologically based differences in behavior, and at least raised the question of whether there might also be ethnically based ones. Feminists, minorities, and other critics from the "nurture" end of the nature/nurture debate argued that Wilson was too biologically deterministic and that he slighted the contributions of learning and socialization to human behavior.

Psychologists were slower yet than biologists to see the potential of evolutionary theory for understanding human behavior. The sole early exception was William James who wrote of instincts as the primary building block of human, as well as animal, behavior. While James believed we could spot animal instincts easily, he said we are nearly blind to our own. Common men, and most academics, were unable to ask: "Why do we smile, when pleased, and not scowl?" "Why are we unable to talk to a crowd as we talk to a single friend?" or "Why does a particular maiden turn our wits so upside-down?" We don't stop to imagine that things could be otherwise. James advocated that psychologists train themselves to "make the ordinary seem strange" and ask the why of any instinctive human act. This phrase perfectly sums up the distinctive perspective that seems to have come naturally to Niko Tinbergen.

Evolutionary Psychology

Each of our ancestors was, in effect, on a camping trip that lasted an entire lifetime, and this way of life endured for most of the last 10 million years.

—Leda Cosmides and John Tooby in *Evolutionary Psychology: A Primer*

The major push to incorporate Darwin into psychology has come under the term "evolutionary psychology." In their primer, Leda Cosmides and John Tooby are fond of saying, "Our modern skulls house a stone age mind." Evolutionary psychologists view the brain as a biological computer with circuits that evolved to solve problems faced by humans and prehuman ancestors. Cosmides and Tooby point out that consciousness is a small portion of the contents and processes of the mind. They describe how conscious experience can mislead individuals to believe that their thoughts are simpler than they actually are. Most problems experienced as easy to solve are actually very complex and are driven and supported by elaborate brain circuitry.

Evolutionary psychologists argue that this is *not* just another swing of the nature/nurture pendulum. They are not stating as boldly as Tinbergen did in his landmark talk that "Nature Is Stronger than Nurture." Their position instead is that it is a false dichotomy: *more nature allows more nurture*. In evolutionary psychology, "learning" is not an explanation—it is a phenomenon *that requires explanation*. They use the example of a larger brained elephant not being able to learn English, not because its brain is less complicated nor even less learning-disposed: elephants have many aspects of memory better than ours. Rather we have evolved specific neural circuits that enable certain types of communication that the elephant has not evolved. Cosmides and Tooby defined evolutionary psychology to exclude the study of individual differences, relegating that to behavioral genetics.

Though sociobiologists and evolutionary psychologists have incorporated many of Tinbergen and Lorenz's ideas, they have not used the concept of supernormal stimuli. Yet I believe it is the single most valuable way that ethology can help us understand the problems of modern civilization. Thus this book examines a range of human dilemmas from the standpoint of supernormal stimuli, interweaving other relevant concepts from evolutionary disciplines to point a path out of our modern dilemmas.

3

Sex for Dummies

In Kurt Vonnegut's novel, *God Bless You, Mr. Rosewater*, Fred Rosewater, a mild-mannered accountant, longs for approval from Harry, a fisherman and a man's man:

> [Harry] scowled at a picture of a French girl in a bikini. Fred, understanding that he seemed a bleak, sexless person to Harry, tried to prove that Harry had him wrong. He nudged Harry, man-to-man. "Like that, Harry?" he asked.
>
> "Like what?"
>
> "The girl there."
>
> "That's not a girl. That's a piece of paper."
>
> "Looks like a girl to *me*." Fred Rosewater leered.
>
> "Then you're easily fooled," said Harry. "It's done with ink on a piece of paper. That girl isn't lying there

on the counter. She's thousands of miles away, doesn't even know we're alive. If this were a real girl, all I'd have to do for a living would be to stay home and cut out pictures of big fish."[1]

Sex is an obvious target for supernormal stimuli. Harry's unusually aware of artifice; most creatures are "easily fooled" as we've seen with Tinbergen's wingless cardboard butterflies and round-bellied wooden fish. Sex shop inflatable dolls—the butt of innumerable jokes and almost as many purchases—are the most literal human equivalents. However, everything from pornography to advertising models, from plastic surgery to old-fashioned cosmetics, cinched waists, and padded bras can be seen as people amplifying nature's signaling. The supernormal stimuli we create tell us much about underlying sexual instincts. They're often designed for one gender, so they highlight differing instincts for men versus women more starkly than do real-world interactions. For that reason, the issues in this chapter apply most directly to heterosexuals, but I also examine what ethology may have to say about homosexuality.

Wanker Nation

Saying that pornography creates a desire for . . . trashy sex is like saying McDonald's creates a desire for salty, greasy meat. Hugh Hefner did not invent the American fetish for women with large breasts; his Playmate of the month merely exploited a taste already well-established.

—Joseph Slade, *Pornography in America*

When virtual reality appeared, pundits warned of the dangers of its sexual applications. One book argued that men—long the gender reputed to be more driven about sex—would no longer require the participation of women. They would soon find their instincts better gratified by computerized virtual worlds. Women, with their need for interpersonal contact, would be left in the cold.

This hasn't come to pass so far. Existing "virtual worlds" are hardly full-fledged multisensory simulations. Rather, they are chat rooms with "avatars" moving around in visual scenery. People have tested the sexual applications of these settings, but they prove cartoonish. As one reporter wrote about the two major competitors: "While scrappy Sims players can build, share and apply homemade nude 'skins' to characters and have them rub together like kids play with Barbie and Ken, Second Life . . . allows players to endow avatars with genitals—I found mine good mostly for laughs in the office."[2]

The Internet has made porn easier to obtain in staggering variety, but it has stuck with tried-and-true formats—still photos and videos. Lower prices for telephone technology have modestly increased commercial phone sex. The occasional efforts at computerized hand-jobs or scratch-and-sniff pornography barely rate mention as a novelty. Humans are overwhelmingly visual, as opposed to many scent-driven animals, and tactile stimulation can be efficiently generated by oneself. (The vibrator is the main technological advance there.) There's not been much effort to combine modalities—no full-fledged supernormal sexual stimuli that is consistently chosen over the real thing.

Of course, old-fashioned porn already has many elements of a supernormal stimulus. At least occasionally, probably much more than occasionally, men choose to masturbate to porn when a real-life partner is available. Proponents say it's "normal for men to look at pornography." If "normal" means average behavior for people in our time, it is—only 11 percent of American men deny ever looking at porn.[3] But if it refers to what the instinct developed for, then it's "normal" to look with interest at an attractive person of the opposite sex, but not, as Harry observes in the Vonnegut novel, to stare at ink on paper.

Does pornography do any harm? At the very least, it's worth

noting that our words for the response to porn—"masturbating," "wanking," "jerking off"—all have the additional metaphoric meaning of wasting time or energy. Masturbation and pornography are not synonymous, of course—most women and some men report that they masturbate with fantasy rather than pornography. But men who describe masturbation as a problem—too frequent, indulged in at inappropriate times, interfering with work or relationships—nearly all report compulsively viewing pornography. A growing percentage of people attending "sexual addiction" programs seek help with pornography rather than problem behavior with real partners.

Other alleged harmful effects of pornography are not supported by evidence. Opponents argue that by objectifying women or by overt representation of rape, pornography promotes sexual violence. However, numerous studies comparing cultures with more or less porn, and tracking ours as porn became more available, fail to find a relationship.[4] In Japan, pornography was decriminalized and became increasingly widespread from 1972 to 1995. The incidence of rape progressively declined from 4677 reported cases in 1972 to the 1995 incidence of 1500 cases; a reduction of two-thirds.[5] At the start of that time many of the rapes in Japan were gang rapes but this has become much rarer so that the number of rapists has dropped even more than the number of rapes. In most of Europe, pornography increased and rapes declined or remained the same through the same period.[6]

Even in countries where violent pornography is allowed, it doesn't seem to increase real sexual violence: Japan has the highest rate of depiction of rape in porn but less actual rape than most of the West.[7] Pornography is also not as violent as its critics usually claim. The largest survey to date found that violence is rare and rape is an uncommon theme in hard-core pornography—*in fact far less prevalent than in mainstream studio films of the same periods.*[8] Even in eras when laws against porn made it such that depictions of rape or child abuse did not add to penalties, these themes were quite rare. Apparently they're just not instinctively appealing to many people. Despite an interest in viewing violence, which we discuss in Chapter 5, most people do not find that violence and sex enhance each other. In Gershorn Legman's classic *Love and Death: A Study in Censorship*,[9] he found that when societies or genres censored sexual images, violent images increased proportionately.

The urban myth of the "snuff" film is an extreme example of this. In 1976, American filmmakers edited and dubbed an old Argentine film that was one of the few that blended the typical bloodfest horror film with pornography. They retitled the film *Snuff*. It featured completely simulated violence—poorly simulated by standards of Hollywood at that time—but the grainy, incoherent film attracted the attention of feminists who picketed its showings, which were otherwise poorly attended. Shortly after, the urban myth bearing that film's title arose, claiming that films of actual murders routinely circulated underground. Despite investigations into this which unearthed absolutely no evidence that this genre ever existed, remarkably many people still believe in "snuff" films.[10]

This is not to say that horrifyingly violent porn images are not out there—of course among the billions of pornographic images, they're easy to find—but so is porn featuring every kind of animal, outfit, or behavior that the average person would find yucky, giggle-worthy, or completely mystifying as to its sexual implications. What the evidence shows is that the majority of pornography is like what Fred Rosewater showed Harry: nude or scantily clad young women with exaggerated breast size, facial prettiness, and enthusiasm for casual sex. At most, the harm is what we see with Harry—wasting time, energy, potential for real relationships. This is, of course, not insignificant in a world where time is at a premium and many people feel a lack of social bonds.

The Fairer Sex

Two types of media for women serve as supernormal stimuli: (1) Images and advice on becoming ideally appealing and (2) romance novels, soap operas, and other media providing vicarious relationships.

"Cosmo" models look suspiciously like male magazine centerfolds. Here, however, they are icons not for anonymous, casual sex, but instead for becoming the one irresistible date and mate choice. We hear how magazines like "Cosmo" hurt female self-esteem with their emphasis on beauty—as if the media had selected this goal at random and taught it to adolescent girls. Or critics suggest that media is doing this to push a commercial, capitalist agenda—a variation on the same fallacy. What sells is hardly random to biology. Anything that sells spectacularly well is probably some type of supernormal stimulus. The media commercially exploit these instincts, but they didn't create them. Humans have always evaluated their personal attributes against those of their neighbors and, for women, physical appearance has always been a large part of these comparisons.

What is different now is that the pool of people for potential comparisons has grown phenomenally—only the unusually attractive are conveyed by media around the world. If a Stone Age girl wasn't the prettiest in her small tribe, the difference wasn't likely to be dramatic. Everyone had opportunities to see others looking their worst—tired, bedraggled, sick—as well as on their best days. Now society culls from millions of young women to select the best faces and bodies, and then perfects these with Adobe Photoshop. The difference between the resulting magazine cover and our average modern girl is staggering. There was a huge outcry when the NBC *Today Show* digitally shaved a bit off Katie Couric's waist in publicity stills and CBS's *The View* did the same with Barbara Walters. But the industry was mystified over the flap: almost all images in our magazines are now digitally manipulated one way or another.

Though the same critiques promote the idea that beauty is culturally determined, the variation in ideal across time and culture is actually modest. Researchers have found that people from different cultures rate the same faces as most attractive.[11] The ratings correlate with a few surface details like smoothness of skin, but most are facial proportions; these follow several general principles. To a certain extent, ironically, attractiveness is averageness.

This fact was stumbled upon a century ago by Francis Galton, the eccentric cousin of Charles Darwin. Galton superimposed the faces of many criminals, seeking to find the criminal face "type." Instead, he discovered that this composite produced a face better looking than that of individual criminals—or any other individual![12] It's partly that composites are more symmetrical, and symmetry itself is consistently found to be attractive.[13] But for many traits we like average dimensions—it makes sense that health lies in the middle of outliers.

On just a few traits, exaggerations are preferred. These are all characteristics that estrogen and testosterone influence. In addition to health, instincts guide us toward the clearest example of the opposite gender and the most fertile ones. In men's faces, this means a stronger jaw and chin. For women, fuller lips, prominent cheekbones, and a smaller jaw. For both groups, babyfacedness is attractive—large eyes, small nose, etc. One study found that the most beautiful female face was one that averaged many women and then blended them 70 percent, with the characteristics of a child's face making up the other 30 percent of the composite.[14] The only exception to childlike features as attractive is where they conflict with the gender-identified ones like a large jaw for men or prominent cheekbones for women.

Body proportions follow similar patterns. Women with a waist-to-hip ratio of .7 are preferred in everything from Greek sculpture to Miss America competitions. Weight has also remained much more standard than is often claimed in modern diatribes, basically ranging from a body mass index (BMI) of 18.5 to 21 in most societies.[15] As the average weight has diverged further and further from this ideal, it's become the most likely way in which two people will vary in attractiveness.

Other traits have receded in emphasis as they became more uniform. When smallpox and other skin infections were rampant and many people were covered with sores or pocked scars, "smooth" or "unblemished" skin "like a newborn babe's" was often included in descriptions of beauty and "slender" was less common. Teeth may be a plus now if they're exceptionally

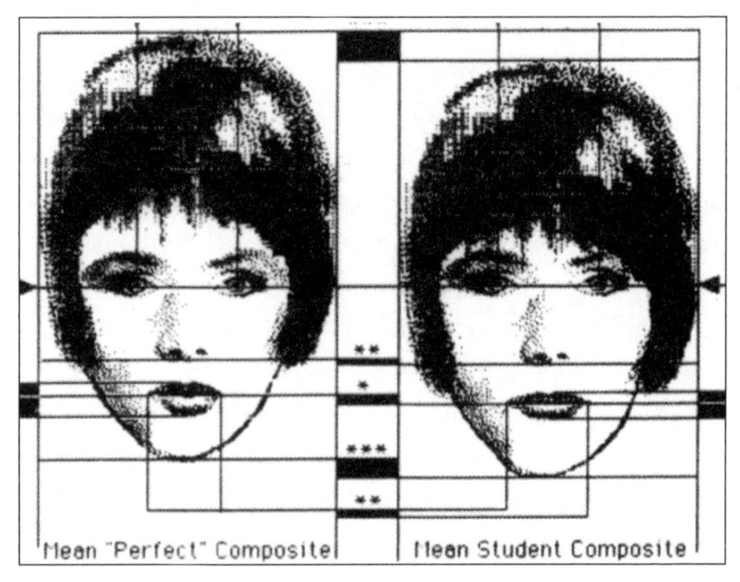

The composite average of many students' faces on the right was rated more beautiful than any individual face. However, altering the upper face to be larger and the jaw smaller—proportions characteristic of children's faces—yielded the composite at left, which was rated most beautiful of all. *Source: V. S. Johnston and M. Franklin, "Is Beauty in the Eye of the Beholder?" Ethology and Sociobiology 14, no. 3 (1993), pp. 183–99.*

straight and white, but the declaration that "I have all my own teeth," once a serious asset for a mate, is currently used mostly in jest.

Before Photoshop and airbrushing, cosmetics offered subtle ways to alter real faces and bodies. Even in Cleopatra's time, women were blushing their cheeks, reddening lips, and drawing lines to mimic larger eyes. Cosmetics provide either modest changes—women with lipstick and mascara are rated slightly more attractive—or none: creams that claim to reduce wrinkles or cellulite merely reduce one's bank balance. Bodies have long been altered by clothing such as padded bras and corsets. Now plastic surgery offers more expensive, riskier, but also more effective versions of these changes for face and body.

Of course women shouldn't base their self-esteem predominantly on their looks, but the tendency to compare oneself to others is ancient. We've developed the ability to make such powerful supernormal attractiveness stimuli and flash them around the world that we are unlikely to ever roll back these idealized images. But it's not useful to deny their innate power, to claim that they are some sort of arbitrary brainwashing. A realistic goal is to remind ourselves that they are an *ideal* only. We manage to find ideals of athleticism, courage, or genius to be inspirational rather than demoralizing. We need to remember that a wide range of appearances is already attractive to our fellow human beings. We also want to make ourselves aware that the ideal is artificially produced and that advertised products don't actually make us look more like the digitally altered images accompanying them. We should promote healthy ways of striving toward (not expecting to absolutely reach) the ideal. "Beauty" advice that actually works includes: get plenty of sleep, eat vegetables, and exercise. The ideal after all is just our evolutionarily imprinted "picture of health."

Porn for Women

To encounter erotica designed to appeal to the other sex is to gaze into the psychological abyss that separates the sexes.

—Catherine Salmon and Don Symons in *Warrior Lovers*

A recent joke book titled *Porn for Women* consists of pictures of attractive fully clothed men with ingratiating smiles in domestic scenes. Dialogue bubbles over their heads read: "I just vacuumed the whole house," "I hope you like chicken sauteed with tarragon and lemon," and "Let *me* get up with the baby." More seriously, some genres not entirely unrelated to these cartoons have been proposed by social psychologists as the most analogous to "porn for women": romance novels, soap operas, and the romantic films often dubbed "chick flicks."

In the romance novel, the point-of-view character is always the heroine. The plot concerns her finding and capturing the heart of the one right man. Sex may be explicit, implied, or not destined to occur until after a proposal or marriage which constitutes the end of the book.[16] In an analysis of 45 best-selling romance novels,[17] anthropologist April Gorry found that the hero was always older than the heroine—by an average of seven years. (Real husbands average only 3 years older in contemporary America.) Heroes are tall; six-foot-two is the most common specific and "over six feet" a popular generalization. Adjectives used to describe Mr. Right's appearance were, in descending order of frequency: muscular, handsome, strong, large, tanned, masculine, and energetic. Personality descriptors were: bold, calm, confident, and intelligent. The most common responses of the hero to the heroine were to: declare his love, want her more than he has any other woman, sexually desire her, consider her unique, and want to protect her. In the words of Salmon and Symons, "A porn video has almost as many climaxes as it does scenes, but a romance novel has only one climax, the moment when the hero and the heroine declare their mutual love for one another."[18]

The readership of romance novels is as overwhelmingly female as porn's viewers are male. And it's nearly as big a market: over 2000 English-language romance novels are published each year, generating $1.2 billion in sales. That's 55 percent of all paperback books and 39 percent of all fiction.[19]

Of course, romance novels are not the only example of what anthropologist Janice Radway terms "exercises in the imaginative transformation of masculinity to conform with female standards."[20] Daytime television "soaps" are another such genre. Mainstream films and popular music serve many other social impulses (we discuss them in Chapter 5), but they also prove a fertile ground for fantasy romance.

In 2008, when *Washington Post* columnist Liz Kelly wrote about her decades-ago infatuation with Shaun Cassidy, she asked readers about their first celebrity crushes.[21] Two hundred and fifty answers poured into the *Post* Web site, describing infatuations begun as young as age five, though early teens were more common. All but 15 were from females and they described fixations on actors, pop singers, and, with lesser frequency, athletes and politicians. Good looks, especially winning smiles were mentioned—and in the case of singers, nice voices—but most women described personalities as capturing their hearts. They taped pictures of their beloved to their bedroom walls and spent hours fantasizing a relationship.

One Macaulay Culkin fan recalls, "I used to pause 'Home Alone' and kiss the TV, believing that where ever he was, he could feel it. I was 8, he was 10—an older man!" An Elton John admirer wore gigantic glasses and wrote a love letter that was never mailed, but instead found by a sibling and read to the whole family at the dinner table. These first crush objects tend to be young, though not as young as the admirer, and adored for their "sweetness." A number of women described that prepubescent crushes on nice guys were later replaced by assertive "bad boys."

Despite the request for "first times," respondents volunteered details about later celebrity crushes. Many of these adult women described still spending many hours in fantasy lives with celebrities. A few even had crushes on the same actors from childhood. One of several admirers of the vampire Barnabas Collins from *Dark Shadows* still watched DVDs from the old TV show. She explained how a not-conventionally-handsome, middle-aged bloodsucker could be an adolescent crush:

> He was soulful . . . a gentleman vampire, a Romantic—suave, sensual, sexy, and sympathetic. He had a wonderful speaking voice (British?), a stalking, seductive style of walking, even a swagger with that cane he carried, beautiful eyes, hair, and hands. He looked aristocratic: haughty, yet haunted, too. He was our Mr. Rochester, Heathcliff, Sydney Carton, Lord Byron, Shelley—all the wonderful literary characters and Romantic poets rolled into one.

It's striking that she invokes romance novels and poetry in explaining vampire appeal. The success of Dracula and King Kong lies in their featuring not only a supernormal alpha male but also a supernatural one. Women ignore the wimpy nominal heroes and respond to the antihero's pursuit of the heroine.

Dracula and Barnabas sleep in coffins, but they're not truly dead as are some other romantic idols. Several of Kelly's repondents' first crushes were on someone deceased before they discovered them. This happened with recently suicided rock singers and long-dead classic film stars—which highlights how artificial our media has made romantic stimuli.

When the men responded to Kelly's query, it was indeed like looking across Salmon and Symon's abyss. One lone male had a romantic reaction to Princess Leia analogous to the women's postings. Other male boomers had crushes on Cheryl Tiegs or Olivia Newton-John because of the "hotness" of their bodies. Younger men debated whether Britney Spears was still crushworthy by describing recent photos of her crotch. A gay man chose an idol favored by many women for a different reason: "David Cassidy after I saw his armpits and the top of his pubes in *Rolling Stone*." Another gay male crush was chosen for "parading his bulge in Speedos." Lesbians who weighed in discussed early Jodie Foster films with the same romanticism other women use for celebrity guys—it's clearly a gender difference, not one of sexual orientation.

Most women easily identify with these crush stories—at least by generation: for the boomers, which Beatle did you like? For the gen Xers, which Duran Duran member? Napoleon Solo versus Illya Kuryakin or Han Solo versus Luke Skywalker? It seems a normal part of growing up—for some, an ongoing part of life even after establishing real relationships. However, just a few generations ago, being in love with someone you'd never met, who lived thousands of miles away, or who'd died before you were born wasn't normal—it wasn't even *possible*: no media existed to support such fantasies. Crush responses were limited to people around you—though probably always stronger for the young and unattached, for women more than for men.

The first large study on romantic crushes was published in 1934 after surveying 350 American high school and college students. Many trends were similar to now. Girls reported more and earlier crushes than boys. Boys described their crushes based on physical characteristics or physical combined with personality while girls described being triggered mostly by personality and intelligence. Neither gender endorsed the third alternative offered: "admired for moral characteristics." Boys' crushes were mostly on their peers; only 1 percent were on women over 25. For girls, slightly older peers prevailed by a small majority; 22 percent had crushes on teachers, and every age was represented. The experimenters had expected, based on existing beliefs among parents and professionals, that many kids who later turned out heterosexual would have had same-sex crushes, but this was not the case. Same-sex crushes were reported mostly by those whose ultimate sexual orientation was homosexual.

One answer from the 1934 respondents was radically different from those in any modern study. Only 3 percent had crushes on "entertainers"—this category wasn't much higher than the number who fancied *monks*. Certainly there have been celebrity crushes since there were media of any kind: Lord Byron was a romantic image for women who knew him only through his books and women flocked to see Valentino films before this survey. But in 1934, film was at low resolution and infrequently viewed, and there were no posters of stars to be plastered on adolescents' walls. Records existed, but not the cult of musicians as personalities to watch on video. There was no TV: people didn't see politicians or athletes up close much. Until recently, most young crushes and ongoing fantasies of adults focused on flesh-and-blood people.

What Is Natural?

Two decades ago, it was common for historians to claim straight-facedly that romantic love is a modern invention. This idea is patently ridiculous to anyone who's ever been in its throes—there's obviously something primal and biological at work. Many animals have observable equivalents. Only in academia could this have been seen as a social fashion. The argument took off from the supposed "fact" that, until recently, women had little say in choosing mates because of arranged marriages. Women were bartered between families and tribes. There was pressure for "within-class marriages." Divorce and adultery were prohibited. Women were held to a very low status—in these or any other decisions.

Actually, historical evidence suggests that most of these factors arose within the last 10,000 years—much more recently in most parts of the world.[22] More than a century ago, psychologist Edward Westermarck's *The History of Human Marriage* pointed out that in pre-agricultural tribal societies, females exercised considerable powers of mate choice.[23] The economic and geographic demands of agriculture limit mate selection, because agriculture requires long-term investment in preparing and maintaining a plot of land, and thereby reduces the physical and social mobility that underlay the free choice of partners in hunter-gatherer tribes. Divorce became much harder in an agricultural society. "Whoever elected to leave the marriage left empty-handed," observes anthropologist Helen Fischer. "Neither spouse could dig up half the wheat and relocate."[24]

Many sociobiologists believe that, with the rise of postagricultural and postindustrial society, we are seeing a return to more "natural" ancestral patterns. They include in this more sexual experimentation in adolescence; higher rates of adolescent pregnancy, divorce, and infidelity; more serial monogamy; more single mothers; and higher rates of bisexuality. "These 'new patterns' probably represent the Pleistocene norm, not 'social problems' due to modern atheistic decadence," concludes psychologist Geoffrey Miller. "The sexual freedom and social complexity enjoyed by young people in contemporary urban North America and Europe is probably much more representative of ancestral tribal conditions than the cloistered, oppressive patriarchy of medieval Europe or the lifelong monogamy of the mid-twentieth-century industrial United States."[25]

Clues from our Animal Relatives

Among our closest relatives, the primates, there is a wide variety of sexual behavior. Gibbons spend their life with one mate. Bonobos have a huge variety of sexual partners—of both genders. Since so many human societies have laws organizing monogamous marriages, many social commentators try to place humans in the gibbon camp—but evolutionary psychologists don't. "Talk of why (or whether) humans pair-bond like gibbons strikes me as belonging to the same realm of discourse as talk of why the sea is boiling hot and whether pigs have wings," writes Don Symons in *The Evolution of Human Sexuality.*[26]

What have evolutionary biologists learned about sexual behavior and physical traits within a species? In species where males have multiple sexual partners, the females in these harems

This orchid mimics the crucial characteristics of the female wasp's torso so well that the male wasp thrusts and ejaculates on it, spreading the plant's pollen in the process. Adaptive for the plant, but one might think evolution would select against an insect who wastes his ardor on a flower. As with human males and pornography, it seems ejaculating in response to anything that approximates a fertile female has been a rewarding reproductive strategy.

are always smaller than the male and dependent on them. The almost completely monogamous gibbons average females only 3 percent smaller than males,[27] while polygamous gorilla males are twice the size of females. Humans, with males usually about 20 percent larger than females, fall in the midrange.

What are the cues to a female's roving eye? It turns out that, across the wildly different genitalia within nature, larger testicles in the male are a sure sign that his sperm have to compete with those of other males. If females of a species are only willing to mate with one male, the efficient testicle is just large enough to supply sufficient sperm for fertilization. When females are inclined to mate with multiple males, the guys evolve larger testes that compete to produce the most sperm. Bonobo chimpanzees which enjoy an orgiastic variety of sexual partners have testes .3 percent of their body weight. Gorillas, whose females mate only with the dominant "gray-back" of their group, have modest testes only .02 percent of their weight. And human males? .08 percent of body weight—four times that of gorillas but one-fourth that of chimps—again suggesting an intermediate pattern.

Research from early times and tribal societies indicates that humans overwhelmingly pair-bond, but with some men having multiple spouses. DNA studies across a wide variety of cultures find about 10 percent of children are not biologically the offspring of their socially identified father. This also turns out to be the case with many birds and other animals previously assumed monogamous by observation until DNA revealed that chicks were sometimes fathered by neighboring males rather than by the one tending their nest.

Some people believe that everyone is naturally bisexual and cite animal relatives such as the bonobo, but actually human cultures across time look much like modern ones—something like 6 percent homosexual. Because hormones in utero play a role in later sexual development, there's been some speculation that hormonelike pollutants in the environment might increase the rate of homosexuality, but that rate appears fairly steady since statistics have been kept.[28]

Sex in the City

Contrary to other patterns which have remained stable, population density is one thing that has changed radically from

Paleolithic norms. Someone living in Manhattan or Tokyo may walk past more attractive possible mates in one city block than their ancestors saw in a lifetime of wandering the savannah. The total number of partners for the most sexually active moderns obviously exceed any in prehistoric times. This translates into higher rates of everything from jealously to venereal diseases. And at least one venereal disease comes from crowding not with other humans, but with *livestock*. Zoo animals confined to unnaturally crowded conditions mate with other species, which they do not do in the wild.[29] Agricultural humans are no exception: syphilis entered the human population from sexual contact with sheep.

Another area of human sexuality about which Paleolithic norms can be instructive is the controversy around the proper age for beginning sexual activity. Within recent history, states in the U.S. have set the "age of consent" everywhere from 12 to 21. One group thinks it's immoral or dangerous to have sex before one is mature and independent. Aside from conservative groups that promote waiting until marriage, most parents and sex educators in America tend to favor 18 or later as the time for beginning sexual activity. However, another contingent thinks it's naïve, unnatural—and probably just plain futile—to try to prevent sex once someone has developed the hormonal drive for it. Holland has a legal code that recognizes sexual consent from the age of 12, but has special provisions for children or parents to bring charges if they can prove adults have used "coercion" on those aged 12 to 16.

Many other European countries and Canada set 14 as the age of consent.[30] Great Britain is presently considering lowering its age of consent from 16 to 14. After the BBC broadcast a program: "Sex Before 16: Why the Law is Failing,"[31] it conducted a phone poll of its viewers. Over 3000 people responded and they were fairly evenly divided between four options: that the age of consent should be reduced to 14, that it should stay at 16, that it should be raised to 18, or that it should be abolished.

Which side is right? Holland's age 12 and "as soon as they feel the urge" or 18 and "not 'til they're truly independent adults"? What does Paleolithic history tell us about this? From an evolutionary perspective, *both* are right. In our ancestral setting, sexual behavior would have begun soon after puberty *and* not until brain development and social judgment were complete. Through most of human history female puberty took place at around 17½ to 18. By 1900, it had fallen to 15½ in the developed world. Over the last few decades, the drop has accelerated until 11 is now the average and many girls reach it at 9 or 10. Early suspects for these changes were sex hormones in milk and pseudo-hormone compounds in the environment. Recent studies show that while these play a role, the single greatest factor is the extra calories in girls' diets and the estrogenic effects of additional fat cells.[32] Boys show a smaller drop in the age of puberty—moving them out of synch with the girls. In their case, growth hormones in the environment speed puberty, but the heaviest boys, because of increased estrogen, are the latest developers.[33]

Meanwhile, medical historians have found evidence that brain maturation may be delayed, though by a more modest one to two years. Full brain maturity that once evolved by 18 may now take until 19 or 20. This also seems to be due to a high-calorie, high-carbohydrate, lower nutrient diet. Excess calories lower levels of brain-derived neurotrophic factor which is necessary for brain maturity.[34]

There's no "natural" answer to the current dilemma of precociously pubertal kids. What we need to do is get growth hormones, estrogens, and sheer excess calories out of our modern diet and environment to bring sex drive and maturity back into synch.

4

Too Cute

Most people don't try to parse cuteness. Like pornography, we know it when we see it. With a bit of examination, however, cuteness has easily quantifiable aesthetics. Take a moment to picture whatever you find cute—puppies, kittens, cartoon characters, or your own children. Cuteness is the type of attractiveness associated with youth, so your "cute" objects no doubt have many youthful traits.

Infants of most species have a small body with a disproportionately large head, big eyes, small nose, chubby limbs, and clumsy coordination. Youthful behavior includes playfulness, affection, helplessness, and a need to be nurtured. A few characteristics such as dimples and baby talk are unique to humans, but most are common across species.

Evolutionary biologists view "cuteness" as simply the mechanism by which infantile features trigger nurturing in

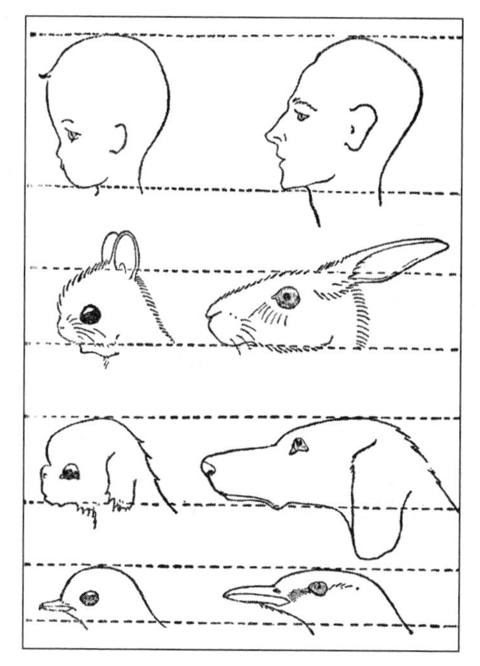

Konrad Lorenz's sketch shows the consistent changes in facial proportions from infancy to adulthood.

adults—a crucial adaptation for survival. Scientific studies find that definitions of cuteness are similar across cultures. So are our responses.

Anyone disheartened by research demonstrating that attractive adults are better liked and better paid than their homelier peers will be further dismayed at studies on infant cuteness. Articles such as "The Infant's Physical Attractiveness: Its Effect on Bonding and Attachment" document that stereotypically cute babies receive the most attention from both strangers and their own parents. They run less risk of abuse or neglect. Cute children proceed to get better treatment from teachers.[1]

"Somehow, Daniel, you're not as cute as you used to be, and you're beginning to lose our attention."

Fortunately, most babies are cute enough to attract sufficient nurturing from parents and the world around them. Mothers and fathers of the "terrible twos" are often heard to remark, "It's lucky he/she's so cute or I would have left him/her on some church steps by now." But, of course, that's the whole point of the features of cuteness and our instinctive reactions to them. The decline of cuteness normally coincides with the child's diminished need for caretaking, which gradually shifts toward younger siblings.

Cuteness signals often go far astray of their intended targets. Humans find baby animals cute and perennially take in foundlings. The animals behave sweetly at first but grow aggressive, literally biting the hand that feeds them. Eventually, they're turned out to zoos, farms, or wild settings in which they're ill-equipped to survive.

Animals can also respond to cues given off by other species' offspring. Recently, a National Geographic TV crew followed a female leopard they named Legadema. They observed her kill a baboon. In the midst of dining on her kill, Legadema spotted a day-old infant baboon on the ground near the carcass of its mother. "The little baboon called out, and we thought we were going to hear a major crunch and the leopard smacking its lips," recalls the photographer. Instead the baby baboon put its paws out and walked toward the leopard. "Legadema paused for a moment, apparently not knowing what to do. Then she gently picked it up in her mouth, holding it by the scruff of its neck and carrying the infant up a tree to keep it safe."[2]

All through that night, the leopard nestled in the tree with the baby, licking it and trying to keep it warm. Several times, the baby baboon fell from the tree. Each time, Legadema raced down before the hyenas picking at the mother's carcass could get the baby and carried it back up to safety. When the sun came up, the tiny baboon was no longer moving. It had succumbed to either the falls or the lack of milk. Successful anecdotes invariably involve an animal who's recently given birth and is lactating. Legadema was not, but she made a valiant maternal effort.

The National Geographic crew decided to make the film about Legadema when they were gripped by similar instincts. They began a project on leopards. "We were filming the adult leopardess when this adorable little cub stuck her head out of the log which was their den," their cameraman recalled. "It was possibly the first time she had ventured into the outside world, and she stumbled around in the sunlight, falling over as if she were drunk." The film became a three-year documentary, *Eye of the Leopard*, following Legadema exclusively as she grew from a cute baby into the predator with maternal instincts toward a baboon.

Occasionally, animals have adopted humans also. Most well-documented cases involve either wolves or wild dogs rearing children, but there has been at least one instance each for bears, monkeys, chimpanzees, and panthers (see Table 1). In 1996, Bello, a two-year-old Nigerian boy, was found after being abandoned at age six months and spending one and a half years with chimpanzees. He walked by bending his legs and dragging his arms on the ground and leapt about, throwing objects. So far, he has learned some human language and social behavior but continues to make chimpanzee-like noises amid his speech.[3]

John Ssebunya was abandoned as a two-year-old in the jungle of Uganda and adopted by a colony of African green ververt monkeys for the next three years. In 1991, a tribe spotted the naked boy. After capturing him, they took him to an orphanage. Ssebunya was more adept at climbing trees than walking and he chattered like the monkeys. He picked up a halting version of human language in subsequent years and enjoys singing in the church choir, but he still lights up most fully when he encounters ververt monkeys. He chatters animatedly with them, apparently understanding and communicating.[4]

One of the best documented early cases occurred in 1920 when Kamala, age 8, and Amala, age 18 months, were discovered in a wolves' den in India. The two girls were probably not sisters, but rather were adopted by the wolves at different

TABLE 1

Children Found Living with Animals 1900–2004

Note: Cases in boldfaced type are described in the text.

Name	Sex	Location	Date	Age (years)	Animals
Andrei Tolstyk	M	Bespalovskoya, Russia	2004	7	dogs
Traian Caldarar	M	Brasov, Romănia	2002	7	dogs
Axel Rivas	M	Talcahuano, Chile	2001	11	dogs
Ivan Mishukov	M	Retova, Russian Federation	1998	6	dogs
Bello	**M**	**Nigeria**	**1996**	**2**	**chimps**
John Ssebunya	**M**	**Uganda**	**1991**	**6**	**monkeys**
Daniel	M	Andes, Perú	1990	12	goats
Saturday Mthiyane	M	Kwazulu-Natal, South Africa	1987	5	monkeys
Robert	M	Uganda	1985	6	monkeys
Sunijit Kumar	M	Fiji	1984	12	chickens
Baby Hospital	F	Sierra Leone	1984	7	monkeys
Kunu Masela	M	Machakos, Kenya	1983	6	dogs
Tissa	M	Tissamaharama, Sri Lanka	1973	11	monkeys
Shamdeo	M	Musafirkhana, Sultanpur, India	1972	4	wolves
Djuma	M	Turkmenistan	1962	7	wolves
Ape-child of Teheran	F	Teheran, Persia (Iran)	1961		apes
Saharan gazelle-boy	M	Rio de Oro, Mauritanie	1960	10	gazelles
Ramu	M	Balrampur, India	1954	7	wolves
Syrian gazelle-boy	M	Syria	1946	15	gazelles
Sidi Mohamed	M	North Africa	1945	15	ostriches
Turkish bear-girl	F	Adana, Türkiye	1937	9	bears
Assicia	F	Liberia	1930s		monkeys
Casamance boy	M	Casamance, Guinea-Bissau	1930s	16	monkeys
Jhansi wolf-boy	M	Jhansi, India	1933	10	wolves

(continued)

Maiwana wolf-boy	M	Maiwana, India	1927		wolves
Jackal-girl	F	Cooch Bahar, India	1923		jackals
Kamala	**F**	**Midnapore, India**	**1920**	**8**	**wolves**
Amala	**F**	**Midnapore, India**	**1920**	**1½**	**wolves**
Indian panther-child	M	India	1920		panthers
Satna wolf-boy	M	Satna, India	1916		wolves
Leopard-boy of Dihungi	M	Dihungi, India	1915	5	leopards
Goongi	F	Naini Lal, Uttar Pradesh, India	1914	14	bears
Mauritanian gazelle-boy	M	Mauritanie	c. 1900		gazelles

times. Both girls walked on all fours, kept nocturnal hours, and showed an aversion to sunlight. They rejected cooked food and would eat only raw meat. Amala died after a year, without learning any human language or gait. Kamala lived eight years and picked up a few individual words. She learned to walk upright but reverted to all fours when in a hurry.[5]

When human children are discovered living with animals, their foster parents invariably mount a valiant fight to prevent their removal. The green ververt monkeys harboring John Ssebunya bombarded villagers with sticks and stones as their charge was dragged off. The wolf pack defended Kamala and Amala so fiercely that the adoptive mother had to be shot. The surviving wolves approached the village repeatedly, howling.

When writing about Kamala and Amala, I found the phrase "in captivity" nearly slipped into my accounts of their deaths. Reading these cases, it's not at all clear that people have done these children a favor by returning them to human society to which they never adjust completely. The bond with the first animals who nurtured them is powerful. Humans do not imprint as rigidly as Lorenz's goslings; we remain more flexible in our attachments. However, a crucial component occurs when young children's cuteness is at its most powerful as a "releaser" for adult nurturing and when children have the strongest drive to imitate whoever's caring for them.

Neoteny

"Neoteny" is a term for preservation of infantile characteristics into adulthood. We find adult humans cuter if they possess either childlike looks or playful, innocent behavior. Criminal defendants with youthful facial features are less likely to be found guilty. If there is incontrovertible evidence of their crimes, however, they receive harsher sentences—as if juries feel deceived.[6]

We also experience a strong pull toward species whose adults are neotenous. Animals such as polar bears are a hit with zoo visitors as cubs, but interest in them slacks off as they reach adulthood. Pandas and koalas continue to draw crowds as big-eyed, chubby-bodied adults. Seals and otters are similarly popular—in this case because their playful behavior resembles the young of other species rather than for any youthful physiognomy.

The widely perceived cuteness of pets such as dogs, cats, hamsters, and gerbils is largely due to their neotenous charac-

"The bunny did not get the job because the bunny is cute. The bunny got the job because the bunny knows WordPerfect."

teristics including friendly, playful behavior, larger eyes, and shorter snouts. Even barnyard animals are cuter than their wild cousins. They're less aggressive and sometimes have shorter faces and/or chubbier, clumsier physiques.

Domestication has brought strikingly similar changes in appearance and behavior to a wide range of mammals—herbivores and predators, large and small. Biologists assume that friendliness and cuteness have been selectively bred for, though not always consciously. Another set of changes are of more practical benefit to humans, such as loss of seasonal rhythm of reproduction. Wild animals in middle latitudes are genetically programmed to mate once a year, during mating seasons cued by changes in daylight. Domestic animals at the same latitudes can mate and bear young multiple times a year and in any season. They reach sexual maturity earlier and produce more offspring per litter than their wild counterparts. This benefits the human owner whether the animals are being raised for eggs, meat, or fur. These reproductive traits have been intentionally selected by farmers.

Another set of characteristics has less obvious causes. Many domestic species have shorter tails or carry their tails curled up instead of long and straight as in the wild form. Domesticated animals may be spotted or piebald while others are white. Hair can turn wavy or curly, as it has done in Astrakhan sheep, poodles, domestic donkeys, horses, pigs, goats, laboratory mice, and guinea pigs. Some animals' hair became longer as with Angora rabbits, teddy bear hamsters, and Persian cats. Many domestic animals are miniatures of the wild ones—though a few are larger. One of the most pervasive traits of domestication is that ears become floppy. Darwin noted in *On the Origin of Species* that "not a single domestic animal can be named which has not in some country drooping ears"—a feature not found in any wild animal except the elephant.

Why did these traits change? It may simply be that animals don't need their camouflage coloring or keen hearing once in captivity and they mutate randomly in variations that would die off in the wild. Or it may be that floppy ears, spotted coats, and long hair are seen as cute by humans and subtly selected for. Floppy ears and soft fur are traits of the young in most animals—though not human young. Later in the chapter, I discuss how some researchers are trying to sort this question out, but first let's review some data from our own doorsteps.

Man's Best Friend

Dogs are the ultimate domestic animal. They have a unique relationship to humans. Other animals were captured and selectively bred, beginning with sheep about 9000 BC and continuing with goats (8000 BC), pigs and cattle (7000 BC), horses (3000 BC), and poultry (1000 BC).[7] These were raised solely for meat at first, so they were bred for tameness and tastiness. Only later did man discover riding horses, shearing sheep, milking cows, and gathering eggs from chickens. This resulted in breeding horses larger (their ancestors are pony sized), sheep fleecier, and so on.

Our relationship with wolves goes back further and may have begun in a more mutually determined way. About 20,000 BC, herds of large prey roamed the last Ice Age landscape and the main hunters were wolves and men. Each had advantages. Wolf packs could chase game faster, circle, and trap it better than humans. Wolves took a bigger chance when they moved in for the kill, however: they could be injured by their prey's sharp teeth and hooves. They had no chance with the largest animals such as mammoths. Humans, with their newly minted spears and arrows, could stand back from the game and finish off animals wolves had surrounded. There is evidence that humans and wolf packs hunted together cooperatively across Africa, Europe, and Asia from 20,000 to 15,000 BC. Humans who liked wolves were likelier to live to reproduce just as were wolves who liked humans.

By 10,000 BC, man was beginning to turn wolves into domestic dogs. A fossil with smaller jaws than wolves' was found at a human archeological site in Iraq dated to that time. This adaptation would not be helpful for hunting, but this "fertile crescent" was the first place man began to settle and farm. Humans began to use dogs to control other animals rather than to kill them.

Several centuries later, images in Egyptian paintings,

Assyrian sculptures, and Roman mosaics depict dogs of many different shapes and sizes. Some emerged that were solely pets: a dog very like the present-day Pekingese (a quintessentially unwolflike creature) existed in China by the first century AD. At the same time, Roman ladies kept lap dogs; their warmth was believed to be a cure for stomach aches—and they definitely lured fleas away from people. A Roman writer of the period gives practical reasons for selecting the color of a dog: shepherds' dogs should be white to distinguish them from wolves in the dark, but a farmyard dog should have a black coat to frighten away thieves.

Once dogs were assigned specific tasks—retrievers, sheepherders, guards, and purely pets—breeds took on different traits. Retrievers love to swim while wolves abhor water. Some guard dogs are larger than wolves but lap dogs are, of course, petite. Northern dogs have heavy coats. There are also consistent traits that diverse dog breeds possess. Some are cuteness factors: dogs have skulls that are broad for their length, larger eyes, and most have floppier ears than wolves. Many behavioral changes such as whining, barking, and submissiveness are neotenous characteristics that wolves show as pups but then outgrow.

Dogs were given their own species designation (*canis familiarus*) because early biologists were influenced by their drastic external differences from wolves (*canis lupus*). Dogs can still interbreed easily with wolves, however, and are essentially a neotenous subspecies of wolf. The transition from wolf to dog occurred over a span of 12,000 years in a haphazard and only semi-intentional manner. Recently, another species has been bred to deliberately re-create the process.

Tame Foxes

In Stalin's Russia, Darwin and his theories were disregarded in favor of the "peasant geneticist" Trofim Lysenko, who propounded a variation on Lamarckian inheritability of acquired characteristics. Environmental influence on traits that could then be passed on to successive generations appealed to Marxists because of the implications for human society. Lysenko advocated techniques such as cooling seeds before planting to ensure that they would grow in cold climes—even yield winter harvests, he claimed. Not surprisingly, results eluded him. His ill-informed agricultural policies damaged Soviet crop yields such that they have only recently recovered. The University of Moscow's biology department was dominated by his goofy experiments. Several opposing Darwinian geneticists were executed while most were exiled to Siberia.

Siberia was not the uniformly bleak prison that westerners, or even other Russians, often imagined. It indeed was cold, and some parts had forced labor camps, but largely it served as an isolation chamber for the generation's most interesting intellectuals, artists, and liberal political activists, to protect the Russian masses from their influence. As a friend who grew up there told me, "In Siberia, the government didn't try to enforce rules. There weren't even that many officials there. Siberia was one of the most liberal societies on the planet."

Siberian universities and research labs were underfunded, but relatively free of political pressures—and well staffed with remarkable faculty. Such was the Institute of Cytology and

Genetics in Novosibirsk, Siberia. Its founder, Dmitry Belyaev, had a special interest in the patterns of changes observed in domesticated animals. He'd been exiled from The University of Moscow for taking a Darwinian approach to studying domestic traits such as appearance and mating season observed across species. He disagreed, however, with the prevailing view among Darwinians that each of the traits had been selected specifically, and separately, in each species. He hypothesized that one trait was crucial and drove the rest.

Belyaev believed tameness and absence of aggression toward humans determined how well animals adapted to life among human beings. Because behavior is rooted in the systems that govern the body's hormones and neurochemicals, those changes, in turn, could have far-reaching effects on the development of the animals themselves. These effects might explain the other observed changes in appearance and reproductive behavior that were consistent across domesticated animals.

Belyaev designed a simple experiment to replay the challenge of domestication and test his hypothesis. He took a single species of wild animal and selectively bred it for one factor: tameness. The animal he selected was social by nature so as to be conducive to taming, a close relative of the wolf but never before domesticated, and already prolific in Siberia—*Vulpes vulpes*, or the silver fox.

The experiment was audacious. Many biologists doubted that Belyaev would get results in an observable period of time. But he was well versed in the selective breeding of agriculture and believed in the power of his one selected trait, tameness.

Belyaev acquired hundreds of fox pups from a fur farm and housed them away from human contact. Once a month, they were tested by an experimenter offering food from his hand while trying to stroke and handle the animals. Foxes that fled or bit when stroked were returned to the fur farm. Foxes that let themselves be petted and handled but showed no emotionally friendly response to experimenters were also dropped. Only foxes who were friendly to experimenters, wagging their tails and whining, were bred for the next generation of Belyaev's research.

Each generation was retested and each yielded larger proportions of friendly foxes. By the sixth generation, there were so many friendly pups, that Belyaev added another category: the "domesticated elite" were eager to establish human contact, whimpered to attract attention, and licked experimenters. Only elite foxes were then bred. Now, 50 years and 50,000 foxes into the experiment—and 20 years after Belyaev's death—more than 80 percent of the pups are born with elite tameness, as eager to please as a dog.

As Belyaev predicted, other changes appeared with the tameness, though they hadn't been selected for. The tame foxes have white patches on their fur, floppy ears, rolled tails, shorter snouts, and smaller skulls. They lost their distinctive musky smell—unpleasant to the human nose. They mature a month earlier than wild foxes. Their reproductive period is somewhat more flexible, though this is the one trait that has not yet equaled other domestic animals.[8]

The results appear to support Belyaev's theory that other traits of domestic animals are linked to tameness. His successor, Lyudmila N. Trut, has identified a physiological basis for the relationship.[9] The domestic fox pups open their eyes earlier, react to sounds earlier, and reach puberty earlier, but they display a later surge in corticosteroids—basically, stress hormones. Because of this, their instinctive fear response appears later and never becomes as pronounced. This combination allows them to interact with humans earlier and to attach without developing fear. Pigment cells migrate so late as to sometimes be nonfunctional. The same biological bases for neoteny may determine why the ear cartilage doesn't stiffen in some of the foxes.

Human Neoteny

This association of cute appearance with neotenous behavior goes a long way toward explaining a hallmark characteristic of humans. Not only are some humans cuter than others, but our whole species may also be especially adorable compared to our primate relatives. This parallels the evolution of other domestic animals. Our childlike traits were first pointed out in 1930 by University of London professor Sir Gavin de Beer in his book *Embryology and Evolution*. De Beer listed characteristics of adult humans that resembled the young of other apes including a rounded skull with absence of a brow ridge, a flat face, and a head that is upright rather than angling forward. Most humans retain lighter skin, with which primates are born, and grow hair chiefly on their scalp, eyebrows, borders of the eyelids, and chin—precisely those places where hair first appears on fetal and newborn primates.

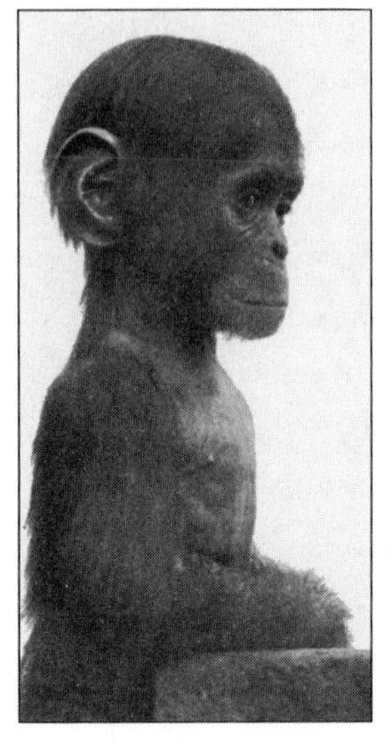

A baby chimpanzee (*left*) has facial and body proportions that more closely resemble those that humans retain throughout life than does the adult chimp (*right*).

More important, we maintain growth and flexibility of our brains much longer than do apes. The brain of a rhesus monkey is 65 percent of its adult size at birth. That of a chimpanzee is 40 percent but we are born with only 23 percent of our eventual brain size. The brain grows the most, but at birth, humans are relatively underdeveloped in many ways. We spend 30 percent of our life span growing—almost unprecedented in the animal kingdom.

As we saw with foxes, neonatal physical appearance tends to go with preservation of childlike behavior. Many aspects of development never stop in man as they do in other animals. Play is seen mostly in immature mammals but adult humans continue to play. Learning of all kinds continues throughout our life span. We've used neoteny in the service of our unique intellect—an aspect we examine in more detail in Chapter 9 on supernormal *intellectual* stimuli.

Cuteness and the Consumer

Walt Disney is said to have pinned a note over each of his animators' desks: "Keep it cute!" His staff not only *kept* it cute, they made it cuter and cuter. As Stephen Jay Gould has pointed out in his essay "A Biological Homage to Mickey Mouse,"[10] Disney's star character became more juvenile in appearance as the years passed. His eyes grew larger, his jaw shrank, and the dome of his head ballooned. Mickey's arms, legs, and snout thickened, his legs jointed, and his ears moved back. By the time the cutification was complete, the adult Mickey looked no older than his little nephew Marty had at the start of the series. Mickey's behavior changed also; in early 'toons, he had pinched, poked, and dominated other animals and maintained a somewhat sexual relationship with Minnie. As he got physically cuter, Mickey became sweeter and less aggressive—and Disney grew richer.

Lorenz himself had remarked on the trend for dolls to get progressively cuter—first they looked like people, then like children, then like supernormal exaggerations of children. Lorenz thought that the Kewpie doll, popular at the time, represented "the maximum possible exaggeration of the proportions between cranium and face which our perception can tolerate without switching our response from the sweet baby to that elicited by the eerie monster."[11]

Around 1990, the journal *Animal Behaviour* published a

series of articles on a creature not of the wilderness but of the marketplace. "The Evolution of the Teddy Bear"[12] traced the origin to 1900 when President Theodore Roosevelt was photographed in the Rockies, after a hunt, with a brown bear in the background. The early teddies looked like bears—with a low forehead and a long snout. Over the years, the teddy "evolved" to become the cute teddies popular now, teddies with infantile features, including a larger forehead and a shorter snout. "It is obvious that the morphological changes that have occurred in teddies in the short span of a little over 100 years have contributed greatly to their reproductive fitness," observed the authors. "There seem to be teddies all over the place."

With tongue in cheek, but metaphor firmly in mind, animal behaviorists continued publishing on the evolution of the teddy. They pointed out that the changes might be likened to mutation, but are actually closer to "intelligent design," diverting human resources to enable teddies to reproduce at a phenomenal rate.

Since a teddy bear is often a child's first toy, one hypothesis that teddy specialists wanted to test was that they evolved to please infants or young children. Researchers offered four- to eight-year-old children their choice of teddies with adult features or ones with infantile features. The four-year-olds chose the adult-featured bear almost two and a half times more often than the baby-featured bear. Among the older children, six to eight years of age, the babyish teddies were three times more likely to be chosen.[13] This makes perfect sense. Very young children are the only beings immune to cuteness. What good would it do a baby to attach to other babies? It is clearly in the babies' interest to attach to adults.

The Evolution of the Teddy Bear from Adult Bear to Supernormal Neonate. *Left,* During the century after its appearance, the teddy got a larger head, wider set eyes, plumper body, and shorter limbs, increasing its cuteness. *Right,* The old-fashioned Santa Claus delivers an older, adult teddy. Both Teddy and Santa got quite a bit cuter through the same period!

The function of the evolved teddy is to please adults—and older children who are already playing at nurturing. These are the purchasers of toys supposedly bought for infants. And teddies are increasingly bought overtly for adults. Dressed in theme clothing, they are a phenomenon on college campuses. Babies will hold the standard babyish teddy when they're not offered the choice of an adult bear. They attach to anything soft and warm, but it's the tactile resemblance to their mother that draws a very young child to the teddy.

Cuteness is an increasingly conscious marketing tool. Elmo, *The Family Circus*, Furby, Precious Moments, and many other cultural icons and products trade on their big eyes and pouty mouths. Disney characters continue to be popular, and Japanese animé such as Pokémon and Hello Kitty feature stereotypically cute protagonists. Hello Kitty has the ultimate small mouth: no mouth. Cuteness can also be a factor in live action films such as the *March of the Penguins*, where the waddling gait of the chubby birds was a major component of the film's appeal.

Advertising, of course, sells many products with cute mascots. Cuteness is second only to sex as a commercial ploy.

The Land of *Kawaii*

The Japanese call cuteness *kawaii* and they have the most *kawaii* culture on the planet. During my first visit to Tokyo a few years ago, I puzzled at brightly painted metal boxes—featuring cartoon dogs dressed in police hats—on street corners. They were police call boxes. Video parlors populated by adults featured games with furry animals in spacesuits competing at futuristic tasks. Winners got their choice of stuffed animals modeled on the characters in the games. These were so large it was difficult to imagine them disappearing inside the tiny apartments of Tokyo.

A reporter for *Wired* described even stranger encounters with *kawaii*: "I bought yogurt one day that had drawings on the label depicting adorable acidophilus-bacteria guys (in white) chasing evil, horned-but-still-adorable fecal germs (in black) out the end of a winding Chutes and Ladders colon." Her Nippon Air jet was painted with Pokémon characters; when it was delayed, the airline gave every passenger a terry-cloth Pikachu beanie doll. While the Western reporter was unmoved, "The Japanese man in the seat beside me, who does something involving industrial drill bits, . . . unwrapped his Pikachu and set it in his lap, so it appear[ed] to be resting contentedly on his balls."[14]

In Japan, you can get married by people dressed as giant cartoon characters, have sex wearing a condom whose wrapper features Monkichi the Monkey imploring, "Will you protect me?" or make offerings at a Shinto shrine from a Hello Kitty charm bag. Not only local animé characters, but Snoopy, Goofy, and Winnie the Pooh are also licensed for every adult product imaginable.

What gives?

"Childhood, in Japan, is a time when you were given indulgences of all kinds—mostly by your mother, but by society, too," says Boston University anthropology professor Merry White. She points out that America is often described as a "youth culture," or as "worshipping youth," but that this refers to adolescence. Our teenagers are catered to by the social structure and many adults long for that more exciting and less responsibility-laden time. "In Japan, it's childhood, mother, home that is yearned for," White says, "not the wildness of youth."[15]

Another contributor may be Japan's declining birth rate—one of the lowest in the world. As fewer adults have toddlers in the home to nurture, there may be more of this instinct available for the *kawaii* cartoon. Japan also has one of the world's densest populations and its most advanced technology. To the extent that Japan deviates from the rest of the world, it may be essentially a glimpse of our future.

We haven't, at least so far, found pets or created stuffed animals that are cuter than our own offspring. It's not clear how well we might survive such an event should it come to pass. Some birds plagued by the cuckoo do eventually evolve a detector and begin to reject the intruder's egg after centuries of victimization. Later chapters describe progressively more dangerous supernormal stimuli: allowing fast food to become a staple in our diet burgeons into a major health crisis; unleashing our aggressive and defensive instincts evolves into wars that could end human life on the planet; and substituting televised sports and drama for real-world exercise and social activities dulls us.

To date, the adverse effects of hijacking the nurturing instinct are subtler. We seem to have enough responses to cuteness to burn. But we want to know when we're doing so. Do we want to be spending as much as we do on toys? Do we want to buy products 'cause their mascots have big eyes or lisp in commercials?

Taking Japan as an example of future trends for us all, they're one of the first countries to hit the wall with overpopulation. They did well at bringing down the birthrate. They've also developed good health care and other advantages that produce the longest life expectancy in the world. But now one of Japan's foremost challenges is the rising number of elderly—expected to triple by 2030. Combined with the falling birth rate, there is concern about who will care for them. Meanwhile, people are paying for *kawaii* computerized Pocket Pets that scream at regular intervals through the night for "feedings." Logic tells us we should switch some resources from toddlers to seniors when their numbers change. But our instincts don't tell us to care for elders nearly as loudly as they holler at us to cater to anything cute. We can't just slap a Hello Kitty mask on the aged; we need to rely more on our intellects to insure that resources are directed toward people ahead of stuffed animals.

You may want to look at the next big-eyed thing that catches your attention a bit more warily.

5

Foraging in Food Courts

Last year, circus veteran Ward Hall grew bored with retirement and decided to revive his World of Wonders, the last traveling freak show in America. Finding midgets, giants and bearded ladies was no easy task in the era of endocrine treatment. In one category, however, supply had outpaced demand. "The fat man—Howard Huge—he wanted to come out with us," recalls Hall. "But I said, 'Howard, a fat man couldn't sell 10 cents' worth of fried chicken. Everybody in America's fat. Go to any buffet restaurant in America; in an hour, you'll see more fat people than in all the sideshows in history.' "[1]

Disney World recently enlarged rides including their trademark, It's a Small World. Built in 1964, the watercourse allocated 175 pounds for the average adult man, 130 pounds for women, and equivalent weights for children. Forty-three years later, overweight boats got stuck against the bottom. Pas-

Circus sideshows sold postcards of people appearing because of their remarkable weight. None would attract a second glance on today's street corners—except, of course, for their clothing or transportation. Two of the cards read: "Ruby Westwood, Age 13, weighs over 17 stone (238 pounds)" and "Giant Boy, Herman. 10 Years of Age, 178 pounds." By contrast, a modern child, Jessica Gaude, has been in the news for her obesity and her 10,000-calories-a-day diet. She is seven years old, weighs 490 pounds and cannot walk.

Adults who have been overweight enough to attract modern media attention include Americans Francis John Lang, Walter Hudson, Michael Edelman, and Rosalie Bradford, all weighing about 1200 pounds, Jon Minnoch at 1400, and Carol Yager at 1600 pounds. Mexican Manuel Garza weighed over 1200 pounds. Hudson, Edelman, Minnoch, and Yager all died between the ages of 25 and 45 of their obesity. Lang and Bradford reduced their weight to around 300 pounds and regained their ability to walk. Garza has dropped his weight to 700 pounds as this goes to press and hopes to be walking soon.

Note: The *Guinness Book of World Records* takes on the task of verifying these weights. More detail about these people's lives appears in "The World's Heaviest People" at http://www.dimensionsmagazine.com/dimtext/kjn/people/heaviest.htm.

sengers listened to the singsong "It's a small, *small* world . . ." repeating until personnel arrived to escort a few people off, releasing the boat. Inconvenienced riders were compensated with free food tickets.

Now a "Small World" with deeper boat channels and larger seats sits next to the food booths. Disney World is hardly alone. My local Boston baseball stadium has installed new seats because the Red Sox fans who come to sit, eat Fenway franks, and watch paid professionals exercise now average four inches wider than those of 1912.[2] The FAA raised estimated air passenger weights after several crashes were linked to outdated norms.[3] Design engineers have upsized everything from collapse-resistant toilets to double-wide ambulances with crane attachments, to XL coffins[4] and jumbo crematoria.[5] The medical profession has just discovered that with standard-length hypodermic needles, most present "IM" injections—for "intramuscular"—now actually end up in subcutaneous fat deposits, releasing medication slower and less completely.[6] They are debating whether to make needles longer or doses larger.

Some modifications are dictated by safety or comfort, but others are psychological. A common strategy to make heavy people feel slimmer is to expand the world around them. Architectural designers call this "framing." "A person who is big does not want to look big," explains David Sokol, editor of *International Design*. "So if their house is bigger, they'll look of more average proportion."[7]

The pharmaceutical industry is frantically searching for drugs to diminish our appetites for the foodstuffs filling our fridges. So far, however, it has enjoyed greater success with medications helping one to live with a terrible diet. In 2008, for the first time, a majority of Americans were on at least one prescription medication for a chronic medical condition. These drugs mostly treat type 2 diabetes, high blood pressure, and elevated cholesterol—all of which are directly related to overeating.

How did we reach this point? By now, you won't be surprised that I invoke the concept of supernormal stimuli. Fast-food and self-interested advertisers did not create our craving for fats, sugar, and salt, though they exploit them. On the African savannah, we evolved a desire for these substances because they were rare and survival depended on locating a bit of each. Now dummy foods loaded with these substances are as close as the vending machine down the hall. We're basically hunter-gatherers lost in one giant food court.

"Worst Mistake in the History of the Human Race"?

The most dangerous aspect of our modern diet arises from our ability to refine food. This is the link to drug, alcohol, and tobacco addictions. Coca doesn't give South American Indians health problems when they brew or chew it. No one's ruined his life eating poppy seeds. When grapes and grains were fermented lightly and occasionally, they presented a healthy pleasure, not a hazard. Salt, fat, sugar, and starch are not harmful in their natural contexts. It's our modern ability to concentrate things like cocaine, heroin, alcohol—and food components—that turns them into a menace that our body is hardwired to crave.

Long before we began "refining" food or creating "junk

Humans also naturally feed on leafy vegetables and lean game. If we eat too much bread, we too get fat and sick and become easy targets for diabetes, cancer, and heart disease.

food" we took the first big step toward decreasing protein, vitamins, and fiber, while upping empty calories. Historian Jared Diamond's essay "The Worst Mistake in the History of the Human Race" refers to . . . agriculture.

Paleolithic hunter-gatherers ate hundreds of plants and animals, supplying generous quantities of complete proteins and vitamins. Ten thousand years ago, the first farmers switched to a diet determined by which plants were easiest to cultivate, harvest, and store. Three crops—wheat, rice, and corn—came to provide most of the calories consumed. These grains are not "natural" foods. They were bred from grasses that produced small seed clusters into plants bearing large, starchy seeds. Each of these grains is deficient in at least one essential amino acid and high in simple carbohydrates which promote weight gain. Beyond these three crops, most people now eat fewer than 20 other plants.

Protein from domestic animals comes with more fat than that from wild game. Animal milk is higher in fat than human milk and contains different proportions of amino acids. The human lifespan dropped seven years with the introduction of agriculture.[8] With medical advances, it has now regained those years and then some, but this hasn't diminished certain diseases that appeared with agriculture: type 2 diabetes, heart disease, and certain types of cancer.

What we *usually* mean by "refining" has reduced the quality of our diet further while intensifying its addictive potential. For centuries, man occasionally hulled grains but, in the early 1900s, methods for refining food became cheap and virtually all grains began to be stripped of the hulls that contained most of their fiber and much of their protein and vitamins. The fattiest cuts of meat—in which large deposits of pure fat had previously been discarded—were ground up for the hamburger that has become our staple. People had been refining sugar from sugar cane—already an obesity- and diabetes-promoting substance. However, corn became increasingly cheap and high-fructose corn syrup rocketed our sugar consumption upward—and it's worse yet for insulin/glucose metabolism.[9] Cheap vegetable oils were hydrogenated into "transfats" to resemble the consistency of animal fat.

In the last four decades, the fast-food industry has perfected the supernormal stimulus.

McHunters and McGatherers

In a *New York Times* satire, Nicholas Kristof describes spotting a six-foot-five figure in a burka while checking email at a Kandahar Starbucks. The imaginary interview with Osama bin Laden reads in part:

> **Q:** So what's your strategic aim? To kill lots of Americans?
> **A:** No. If we wanted to do that, we'd have our agents open up McDonald's franchises . . .[10]

Fast food is so universally understood to be unhealthy that no one needs the joke explained. McDonald's is the most popular target for such jabs if only because it's the largest, super-size target. In April of 2004, Morgan Spurlock released his documentary film *Super Size Me* in which he gained 25 pounds, 60 points of cholesterol, and a fatty liver in just one month of eating exclusively McDonald's food. Three months later, denying it had anything to do with Spurlock's film, McDonald's announced they were phasing out their "Super-size that?" promotional campaign. In the early 1990s, it ran an even more intensive advertising campaign, the "Big Mac Attack"—a series of vignettes about people overcome by an irresistible urge to consume the product. In 1992, journalists

"Yes, but take away the rodent droppings and the occasional shard of glass, and you've still got a damn fine product."

used the phrase in covering a woman's collapse from anaphylactic shock after a rare allergic reaction to a Big Mac. The term cropped up in numerous cartoons about heart attacks. McDonald's discontinued the Big Mac Attack ads, again denying any relation to the new connotations.

But McDonald's survived—even thrived—in the face of negative publicity. The years 1990 to 1994 saw the "McLibel" trial—the longest running lawsuit in the history of the British legal system. McDonald's brought charges against two Greenpeace activists for distributing a pamphlet called "What's Wrong with McDonald's?" After spending $10 million in legal fees against the two defendants who served as their own counsel, McDonald's won . . . technically. The defendants were found liable for $78,000 in damages because the court judged them not to have proved all of the claims in the pamphlet. The court did deem them to have proved that McDonald's "exploited children" with its ads and endangered the health of its regular customers. The defendants appealed and, in the new hearing, the judge ruled they had also proven that McDonald's food can cause heart disease. Some allegations were still judged unproven and libelous—mainly statements about mistreatment of workers.[11] McDonald's chose to withdraw the suit at this point. Pundits pronounced this a "public relations disaster," but McDonald's continued to increase profits throughout the period—including from its British franchises.

McDonald's customers are perhaps not the only people who should worry about the health effects of burgers and fries. In the spring of 2004, as *Super Size Me* was showing in America's theaters, McDonald's CEO James Cantalupo, age 60, a 30-year veteran of the company and famous for consuming the product in public dropped dead of a heart attack at a McDonald's franchisee convention in Orlando, Florida. Literally within hours, the company appointed a successor: 43-year-old Australian Charlie Bell who'd begun his McDonald's career flipping burgers and mopping floors at age 15 and worked his way up to the top spot. Seven months later, Bell was diagnosed with colon cancer—another disease whose incidence rises with a diet high in fat and red meat and low in fiber. Bell died four months after diagnosis. McDonald's news releases discussed the financial and organizational impact of two successive losses, but studiously avoided any mention of public relations or health implications.

Is McDonald's any worse than the rest of the fast-food chains? No—or only to the extent it's doing more business so it's doing more harm. Since the 1990s, the watchdog group The Center for Science in the Public Interest (CSPI) has designated a monthly "food porn" award. Rarely has McDonald's won. CSPI has recognized Hardee's Monster Thickburger at 1420 calories, The Cheesecake Factory's carrot cake with 1560 calories, and Mrs. Fields' 1070-calorie cinnamon roll.[12] The awards haven't embarrassed restaurants into lowering these numbers: the totals just keep going up. In 2001, CSPI targeted Starbucks' 20-ounce White Chocolate Mocha with 600 calories; by 2002, Starbucks had introduced the Vanilla Crème Frappuccino with 870 calories—half the calories a small, inactive woman should consume in one day contained *in a single cup of coffee*.[13]

It's almost impossible to parody this. In 2005, Mike Judge, creator of *Beavis and Butthead*, directed *Idiocracy*—a film about a goofy dystopian future in which the President is a former professional wrestler and monster-truck driver. The public sits slack-jawed in front of the Violence Channel drinking "Brawndo: The Thirst Mutilator." The highly caffeinated, highly carbonated bright green swill has replaced water, which is now used only in toilets. People irrigate plants with Brawndo. Plants are dying. No one can figure out why, but it doesn't matter because the new food pyramid recommends consuming nothing but Brawndo.

Idiocracy earned only $400,000 at the box office, hardly the smash hit that usually licenses its products. However, one evening in 2008, I walked past a garish display for yet another energy drink at my local convenience store. Its name meant nothing to me, but two of its promotional claims got my attention: "With 5 kinds of sugar. . . . It's what your plants crave." Brawndo had arrived in the real word. When Brawndo's manufacturer, James Kirby, first saw *Idiocracy* he thought, "Based on how things are going, especially our country, this is a shoo-in."[14] Selling well in Massachusetts, it's now distributed around most of the Northeast.

P. T. Barnum said no one ever went broke underestimating

Adding and removing volume is an obvious interest for sculptors, but it's given a sociocritical twist by Austria's Erwin Wurm. In 1993, Wurm wrote an instructional book on how to gain two clothing sizes in eight days. Eight years later, he made *Fat House* (*top*) and *Fat Car* (*bottom*) which he described as "taking on the link between power, wealth and body weight . . . examining our current value system, as the advertising world demands us to stay thin but to consume more and more."

the intelligence of the American public. I'd add: no one ever went broke *over*estimating the power of the supernormal stimulus. These are closely related statements. We humans aren't actually stupid—just prone to following outdated instincts instead of stopping to think things through . . . which is what our giant brains are there for.

Beyond the supernormal pull of sugary, caffeinated drinks and other junk food upon first tasting, repeated exposure triggers long-term biological changes. Kirby's previous beverage product was named "Cocaine." Pulled from the market in 2007 when it "ran into a little problem with the FDA,"[15] its name was even more apt than he may have realized. Caffeine is a drug that substitutes for the coca originally in colas, but drug addiction is also a good model for how our bodies respond to artificially refined carbohydrates, fats, and minerals.

What happens when we eat supernormally stimulating, highly refined foods repeatedly? With simple carbohydrates, glucose levels soar in the bloodstream. In the short term, our bodies release insulin to store the glucose as fat. In the long term, we respond to this high level of insulin with decreasing sensitivity to it—developing diabetes, which damages our kidneys, eyes, and immune system. Possibly the worst aspect of these foods is that we can happily eat a lot before our bodies register that we're full; very soon, as glucose levels plummet, *they trigger more hunger.*

There is growing evidence that sugary foods influence the same brain chemicals affected by addictive drugs. Researchers at Princeton have shown that natural opioids are released when rats eat a large amount of sugar and that they're thrown into a state of anxiety when the sugar is removed. Symptoms included chattering teeth and the shakes—very similar to those seen in people withdrawing from nicotine or morphine. When the rats receive naloxone, a drug that blocks opioid receptors, dopamine levels drop and acetylcholine increases. This is the neurochemical pattern shown by heroin addicts as they go into opioid withdrawal.[16]

Biologists find that overeating refined fatty meals triggers similar physiological changes. Leptin and ghrelin are hormones that signal the body to begin and cease eating. After just a few extremely high-fat meals, rats lose their response to these cues and continue to eat excessively.[17] Fortunately, the effect reverses if they are taken off the high fat for a while.[18]

Junk-Food Junior

Children are gaining weight at an even faster rate than adults. The number of overweight children in America has tripled in the last three decades. This is directly related to an increase in their junk-food consumption. One-third of American children eat at fast-food restaurants every day. These children average 187 calories a day more than other children, which results in an extra six pounds a year assuming equal exercise. And while the children's menus at fast-food restaurants are as unhealthy as the adults', at sit-down chains, the kids' offerings are *worse* than the adults'. The Center for Science in the Public Interest surveyed menus at the largest 20 chain restaurants in the United States including Applebee's, Chili's, Cracker Barrel, Denny's, Olive Garden, and Outback Steakhouse. They found that adult offerings usually included grilled chicken entrées and sides of vegetables and salads, but the children's menus featured mainly fried entrees, accompanied by french fries and a dessert.

And our schools? "7-Elevens with books," is how Yale diet expert Kelly Brownell describes them.[19] Fast-food franchises such as Pizza Hut are manufacturing many of their offerings or even running their lunchrooms. *Super Size Me* dramatically illustrated how children are actually eating at school. Spurlock filmed kids buying combinations of chips, fries, cake, and soft drinks in a school cafeteria line. On camera, the cafeteria director assured him they were fetching side orders for a whole table of friends who'd brought standard entrees from home. But when Spurlock followed the kids to the table . . . guess what? They were indeed making meals of the snacks—if they'd brought foods from home, they were often additional sodas or desserts.

When children bring lunch from home, it's more and more likely to be processed, prepared foods. Perhaps the most popular item for the school lunchbox is Kraft Food's Lunchables. In 1988, when these debuted, all were under 300 calories. By 2003, Kraft was making Mega Pack Lunchables like "Deep Dish Pizza—Extra Cheesy" with 640 to 780 calories.

Overseas

Another group gaining weight faster than American adults is . . . the rest of the world. During my 1973 undergraduate semester studying in Paris, I observed French haute couture shoppers lining up by the hundreds outside the newly opened golden arches on the Champs-Elysees. Twenty years later, I traveled to Kuwait in the wake of the Iraqi occupation to teach Kuwaiti psychologists about trauma therapy and watched their first McDonald's open to similar crowds. When I was in Europe in 2007 giving talks about *Waistland*, Europeans were eager to read and talk about the "American" obesity epidemic. Many told stories of visiting the U.S. and being amazed at the size of people wandering around malls and playgrounds. However, even though we "lead the world" in obesity, our rate of increase has slowed as the fast-food market reaches saturation and the rest of the world's is accelerating as they adopt our supernormal foods.

Fast-food restaurants have perfected the supernormal stimulus—and it's spreading around the world like a virus. In more and more locales, the obese are up against a situation similar to the heroin addict living in a neighborhood with a pusher on every corner.

McDonald's and other Western chains do run into the occasional cross-cultural problem, but when they do, it's an issue of politics rather than menu. When U.S. troops entered Iraq, McDonald's experienced calls for boycotts and suffered numerous acts of vandalism around the Middle East. Most American-based chains had problems—except Pizza Hut, whose relieved CEO speculated, "I guess they think we're Italian."

Some think the popularity of McDonald's, Burger King, and Pizza Hut around the world lies in a glamorous American cache—and there's *some* truth to that. But there's also a deep animosity toward America in much of the world as demonstrated by the attacks on franchises. The phenomenal success lies largely with what McDonald's and other chains are selling.

The major fast-food products are remarkably similar—chains have taken products from disparate countries and shaped them toward identical ratios of refined carbohydrate, saturated fat, and salt—and then served them in huge portions. These are followed by desserts with similar refined carbs and fats but with sugar instead of salt or meat—essentially the same ratios, albeit with much cheaper ingredients, that French pastry has long contained. Fast foods the world over show few variations from the formula, such as the McArabia sandwich—basically a Big Mac on Middle Eastern flatbread. Some Asian fast food still features totally refined rice instead of wheat, but it's mixed into virtually identical portions of fat and either sugar or salt.

In a world increasingly designed to stimulate hunger, "listening to what your body wants" is a losing strategy. It's not antihedonistic to rein in, or redirect, instincts. Our pleasure system is robust and *very* flexible. Our brainpower can direct it—indeed that's what it evolved to do.

Scientific studies show that people experience similar levels of happiness over the long-term regardless of external events. Winning millions in a lottery, or getting paralyzed in an accident, make a significant difference for six months or less.[20] People living in poverty average only a few percentage points less happy than the most affluent, and there is no difference between the middle class and the rich.[21] In fact, the only thing that makes a difference is chronic pain or consistent health crises—things that long-term overeating produces.[22]

At first, swearing off french fries sounds harder than ordering the small size, more difficult than a regimen where weight loss is rewarded with a serving of your favorite dessert. If it's a hardship to go a week without eating a single cookie, then it's natural to assume that it would feel infinitely worse to go five years without one. However, the basic physiology of glucose, leptin, and ghrelin mean that radical changes in diet are often *easier.*

Let's return for a minute to the analogy of drug addiction. We all hear people talk about how they couldn't possibly give up cheeseburgers and fries *entirely* or that it's cruel or unreasonable to suggest they never eat dessert. But as a psychologist, I hear from addicts how completely unimaginable never shooting up again seems or how they just couldn't get through the day without a certain number of drinks or pills. Both are compellingly heartfelt but not entirely accurate. People think I don't really understand how much their boy *loves* ice cream if I suggest they skip it—but he'd *love* heroin if they were doling it out after dinner. The pleasure mechanism can be shaped as to what it responds to—it doesn't have to be the other way around. When you begin to eat healthfully, within days, glucose and hunger-regulating hormones shift, diminishing cravings. Within weeks, a positive conditioned response becomes associated with fish or spinach and extinguishes to french fries or mousse. Hunter-gatherers on the savannah—or health-food enthusiasts in their chi-chi modern restaurants—*really* do enjoy their fish and broccoli as much as the connoisseurs of French pastry or donut shop regulars do their repasts. It's only the continuing consumption of the supernormal stimulus that renders the natural one unappealing. Unlike the complex paths our nurturing, sexual, and romantic instincts have taken, the hijacking of our drive for nourishment clearly needs to be reined back right now toward something more like what our ancestors practiced.

Individual Change

What people should do in our current bloated food environment is no diet secret. It's simple and obvious and known to most: eat lots of vegetables, moderate amounts of lean protein and fruits, and small servings of nuts, seeds (including grains), and eggs. No one should eat *any* transfats, white flour, or refined sugar. Overweight people need to eat fewer calories—there's no magic number: just drop calories until you're losing weight. Exercise.

People often complain that it takes too much time to prepare healthy meals but it takes no time or effort to dump a small can of tuna over half a bag of baby spinach, for example—just the desire to do so. For those who crave detailed lists of preferred foods and recipes for preparing them, many good books exist.[23] However, don't let that effort serve as an excuse to postpone eating healthier.

Since the *what* to do is elementary, most people fall down on the *how*.

HABITS With modern brain-scan technology, scientists can see our "habits" at work. When we do something familiar, an area deep within our brain—the basal ganglia—fires an exact sequence of commands without our consciously having to think about it. This is how we drive or walk our habitual route to work while our mind may ponder other matters. When a route is unfamiliar, a very different brain area—the prefrontal neocortex—lights up as we debate whether to turn left or right, as we decide which street is likely to have more traffic. This mass of gray matter just above our eyes controls the conscious weighing of complex options. It works slowly and carefully. Unlike the basal ganglia, which is shared with other vertebrates, the prefrontal cortex exists only in mammals. It is largest in apes and most of all, man.

As we learn the route, the basal ganglia begin to generate a known pattern—and actually suppress other alternatives. The prefrontal neocortex either quiets or turns to other matters like a task we'll start upon when we arrive or vacation plans next month.

If we've fallen under the sway of a supernormal stimulus, we need to change our habits to avoid it. The first stage of change is to engage the prefrontal neocortex and ponder healthier options. It gives us choices to override instincts and habits, which other animals don't have. When we begin a diet, this area is active with questions like: "What's the healthiest entree on this menu?" or even "Do I really want to stick with the diet or would I rather order some ice cream?" However, we don't want to stay in the frontal area indefinitely. It's effortful, and we can always make the unhealthy choice. We want to perform the best action so consistently that our basal ganglia take over and soon we don't weigh options any longer—the new habit is in place: the waitress approaches and we order the garden vegetables.

PSYCHOTHERAPY Conventional psychotherapy has sometimes offered misguided advice on overriding habits and instincts about diet. Long-term therapy aimed at understanding how one eats to "fill an inner emptiness" as if it is an individual quirk is dubious in an environment when two-thirds are overweight and almost everyone will overeat if dining on supernormal stimuli. There are two psychotherapy approaches, however, that have proven helpful to people losing weight—and to habit change in general. Cognitive behavioral therapy and hypnosis can both be effective if focused specifically on eating and exercise behaviors.

Cognitive behavioral therapists ask clients to keep a log of the problem—for excess weight, charting everything one eats and all exercise. Recording eating makes it less automatic and this alone often decreases intake or spurs healthier choices. Clients are also asked to note *when* they overeat, what external cues were present—components Tinbergen would have observed about gulls or wasps. But cognitive behaviorists also ask clients what their thoughts were. Together they examine automatic thoughts for faulty reasoning and possible better ways of interpreting the situation, inviting the prefrontal cortex to override established habits and beliefs. For instance, every time someone goes to "get coffee" at a donut shop, she may in fact end up with at least one donut or when someone reasons he will eat a large lunch so he can get by on a small supper, but his supper is never actually small.

Unlike with drug abuse, some of the faulty beliefs about weight loss come straight from society around us—even from popularized misinterpretation of scientific research. Many recent books on the "fat gene," "thrifty gene," or "body types" promote the idea that some people are biologically destined to be overweight. The Pima Indians of Arizona are often cited as evidence for this. The Pima have an average BMI of 30.8 for men and 35.5 for women, so the *average* Pima is obese—69 percent fall into this range. Seeing photos of the outsized tribe lumbering around diabetes clinics, it would be easy to imagine they are "built that way."

But, as we discussed in Chapter 1, there is no such thing as a genetic determinant that operates independent of the environment. The Arizona Pima live a sedentary lifestyle and eat lots of highly processed, high-calorie, low-nutrition food. They indeed get heavier on average than other people who eat a similar diet. But they have tintypes of their grandparents—and these people weren't particularly overweight. Furthermore, a group of Pima who stayed behind in the Sierra Madre mountains of Mexico still hunt, gather, and practice a bit of strenuous farming. They're not only thinner than the average American Pima, they're also thinner than the average American.

Some people insist there is no diet on which they'd lose weight. But they'll agree there's a diet at which they'd starve, so it's pretty easy to get people to reason through what would happen halfway between that and their present intake. Also, most people underestimate their food intake. A research tool called "doubly labeled water" can measure how many calories a person has consumed. When this is compared with dieters' self-reports, overweight people report 81 percent of their actual consumption and obese people report 64 percent.[24] So it's helpful to train people in accurate caloric estimation.

Cognitive behavioral therapy substitutes rational, constructive thoughts in place of irrational, self-defeating ones: "I'm so overweight now, I could never get to my ideal weight," becomes "Of course I can, and that makes it more important to start now." Overweight people start off saying, "I need to reward/comfort myself with food," but it's not a reward to take years off your life, diminish your energy and intellectual capacity, depress yourself, and increase the odds of painful diseases. These are things we shouldn't wish on our worst enemies. Clients in cognitive behavioral therapy learn to reward themselves with a paperback novel, a visit with a friend, or a massage. Most important, they begin to define taking care of themselves, eating light, feeling themselves growing healthier as the real rewards. Our prefrontal cortex helps us form these connections once they are pointed out; no other animal can do this.

HYPNOSIS Research shows that hypnosis is effective for weight loss—either by itself or in addition to cognitive behavioral therapy.[25] During hypnosis, I tell my clients they will enjoy low-calorie, healthy foods and feel satisfied by small servings. I suggest that problem foods will begin to look nasty, toxic, unreal, or simply uninteresting. Phrasing that is especially powerful suggests an involuntary component to this: "*You may find to your surprise* that you . . . [just ignore rich foods], [go the whole time between meals without thinking about eating]." People sometimes repeat back to me in the next session something like "to my surprise, I went all afternoon without even thinking about food," seemingly unaware that this was in their instructions.

Hypnosis affords a "sneak preview" of achieving future goals. People can vividly experience stepping on scales and seeing a lower number, noticing clothes zip easier—or their long-term goals: being slim again, having energy for sports, getting a good medical report, knowing they're going to be there to see their grandchild grow up. Some people are more susceptible to hypnotic suggestion than others and, for these high-hypnotizables, the visions of the future are dramatic. Suggestions can make them *really not feel hunger or cravings.* This is not so surprising—hunger can be as powerful as pain, but not more so. These are the same people who would get complete pain relief with only hypnosis as anesthesia. Pain and hunger are both survival signals but they can both be overridden in people of high hypnotizability, so hypnotherapists are careful not to suggest one will never be hungry at all.

For people of moderate hypnotizability, imagery of a positive outcome is motivating even though it doesn't have the same hallucinatory quality. The hypnotist's verbal phrases come to mind at times of temptation to reinforce their goals—the moderate hypnotizables just don't get the same complete free pass on physical cravings as their bodies adjust to changes.

Changing Society

Ten years ago, America's government, press, and public were desperately aware that we were in the grip of an obesity epidemic. Two-thirds of Americans were overweight. Obesity-related illness was killing 300,000 people a year, sickening millions, and costing $99 billion annually in medical costs.[26] Excess weight was the single most common thing people said they disliked about their own body. Two-thirds of Americans ranked losing weight as a goal of moderate to high importance.[27]

In response to this situation, our country has made a number of changes. Americans have eaten 50 percent more fast-food meals and five more pounds of sugar a year. U.S. obesity-related health costs have risen to $117 billion[28] and medical epidemiologists now estimate 8 out of 10 Americans will eventually become overweight.[29] Fewer Americans are dieting than a decade ago—and with less success.[30] In the previous section, I discussed individuals who are in denial or looking for an easy fix. But our whole society and government are also in denial about what it will take to reverse the obesity epidemic, just as we were with smoking a few decades ago. In 2004, the World Health Organization proposed dietary guidelines to reduce fat and sugar consumption. The U.S. delegation—representing the fattest nation in the world—protested this, on behalf of the food industry, as "scientifically unproven."[31] The WHO guidelines call on governments to restrict food advertising aimed at children, to use fiscal and pricing policies to discourage consumption of junk foods, and to pressure the food industry to reduce the use of unhealthful ingredients like trans fatty acids. The U.S. Department of Health and Human Services responded that instead, it would be better to use approaches like "better data and surveillance, and the promotion of sustainable strategies that focus on energy balance," terms so vague as to be meaningless. If the Administration wanted better surveillance about obesity, it should dispatch HHS staff to shopping malls and playgrounds rather than sending them to Switzerland to sabotage WHO antiobesity efforts, responded consumer group CSPI.

There are a few promising initiatives toward controlling foods. A few jurisdictions have passed local laws banning transfats or limiting the percentage foods can contain. Similar proposals have been voted down in other jurisdictions, but it seems to be slowly happening. Limiting junk-food advertising to children and eliminating the worst foods in snack machines and cafeterias passed a few places. But it's hard going. When Congress enacted a law prohibiting schools from selling sugary sodas in school cafeterias, so many schools began to evade it by *giving* soda away with lunch—or even breakfast in one school—that legislators had to vote again to close this loophole.[32]

There are other steps America should begin taking now. The U.S. Department of Agriculture should not oversee so much of our dietary advice. The USDA's core mission is selling agricultural products—and too often that translates into maintaining entrenched interests in unhealthy products. As former senator Peter Fitzgerald noted, putting the USDA in charge of our nutritional guidelines is "like putting the fox in charge of the henhouse." USDA subsidies to farmers for growing specific crops total $19 billion annually. The vast majority goes toward producing such unhealthy foods as white flour, white rice, butter, oils for hydrogenated margarine, and corn for corn syrup. Not a dime goes toward growing broccoli, spinach, farmed salmon—or most foods proven healthy by medical research. No one wants to hurt the American farmer, but we needn't subsidize them at the cost of the public's health. A common excuse for eating processed foods packed with carbohydrates and fat is that fresh vegetables cost more. Subsidizing vegetables would help the farmers who grow them—and all of us.

Another example of the questionable agenda of the USDA is its grading of commodities with categories that lead in exactly the wrong direction. Grades of meat are determined by fat content with "Prime" as the most fatty and "Regular" as the least. If you wanted to direct people toward supernormal stimuli and away from healthy choices, this would be the perfect system.

The USDA's Agricultural Marketing Services generate slogans for products. Its "Everybody needs milk" was ruled by the Federal Trade Commission to be illegally misleading. Since then, they have helped Wendy's develop its Cheddar Lovers' Bacon Cheeseburger as a "cheese-friendly sandwich." Pizza Hut's "Summer of Cheese" featured two recipes developed with help from the board. Why not encourage foods that use more tomatoes, more spinach? Let's put the board to work on slogans for fish, recipes for soy.

Small Business Administration (SBA) loans are another example of perverse government incentives. The loans were established to help entrepreneurs start small businesses, but SBA loans can also be used to open a franchise of a national corporation—including fast-food chains. Most of the loan start-up money goes to the corporation for the franchise rights and their mandated equipment. Many critics of this program want simply to redirect this money to genuinely independent businesses, but why not also require the new businesses to serve a beneficial purpose—or at least to do no harm? Only a restaurant or store selling primarily healthy whole foods would be eligible for a loan. A restaurant to sell cheeseburgers or ice cream wouldn't—nor someone proposing yet another "convenience store" selling donuts, potato chips, and cigarettes.

Health warnings on risky foods are another technique short of an outright ban that we can borrow directly from the anti-smoking campaign. People were not getting the information in 1964 when for the first time their cigarette packs began to state that the Surgeon General had determined smoking caused can-

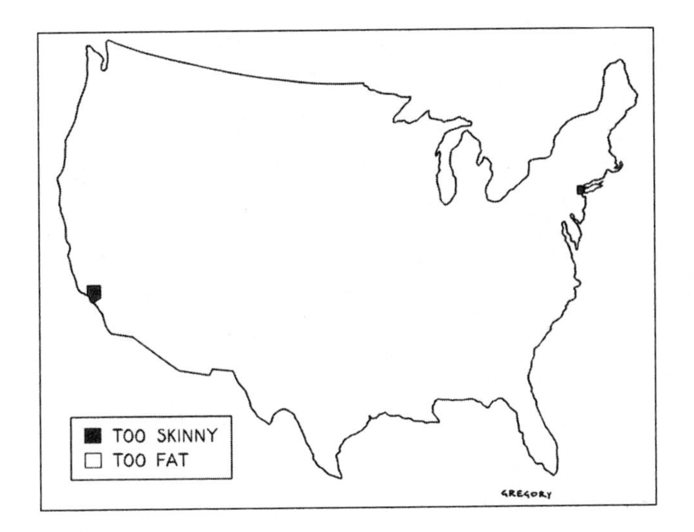

cer and birth defects. But over the years, these warnings gradually discouraged smokers. Similarly, most are aware of the links between sugar and overeating with diabetes, but it might improve eating habits to label some foods: "Eating refined sugar and flour increases your chances of diabetes which can result in kidney failure, blindness, and amputations." On others: "Saturated fats increase the risk of cancer and heart attack," or "Eating more than the recommended daily calories raises your chances of early death and Alzheimer's disease." There are similar initiatives that would structure cities to encourage exercise and school systems to upgrade physical education.

The solutions necessary and practical in the third world differ from those for America and other industrialized countries. Most of the third world still gets vigorous exercise. Their food problems include scarcity—though the WHO recently announced overfeeding is now a wider problem than starvation for the first time in history. And whether over- or underfed, both types of cultures are frequently under*nourished* because they eat simple carbohydrates and saturated fats, short on vitamins. Poorer countries are not as permeated with fast foods. Certain categories, however, such as soft drinks, are even more of a problem. Babies and very young children sometimes get sugary sodas instead of milk for lack of education.

Obviously much of the world can't afford to spend the amount on food that the U.S. or Europe or Japan can. Americans, if they choose, can buy vegetables grown everywhere in the world—more fully replicating the variety available to hunter-gathers. If one has that selection, grains are not the best choice for the bulk of the diet. Ideally, whole grains should be a small fraction of what we eat. But in much of the third world, whole grains supplemented with smaller amounts of vegetables and protein is the only practical, affordable solution that's much better than what's eaten now. At its current population, the planet can't support a hunter-gatherer lifestyle or the sort of agriculture that fully replicates it. In the long term, lowering the birthrate to control population might solve many environmental issues, but this is obviously a many-generational solution.

Getting mankind to stop refining grains is a priority. We must use this staple whole with its fiber, essential oils, and vitamins intact instead of turning it into white flour, white rice, corn syrup, and products manufactured from them. Whole grains need to be supplemented with some vegetables and protein sources—relying on those that are cheap and easy to grow locally. In Bangladesh, beans and pumpkins thrive readily and supply vitamin A and other nutrients. The U.N.'s Food and Agriculture Organization—which is much more health conscious than the USDA—has helped landless families develop home gardens to grow these crops for at least 3 million people. In Thailand, the ivy gourd grows easily and is again a good source of vitamin A and some of the B-vitamins. A project in the Jiangsu province of China initiated rice/aquaculture systems, which resulted in 10 to 15 percent increases in rice yields and, more importantly, 750 kg of fish per hectare of rice paddy. The fish also help reduce the incidence of malaria by consuming mosquito larvae. In many more areas, soy is the cheapest producible healthy protein and can be a lifesaver.

Not everything modern technology has yielded is negative. Green plants and animal protein don't have to be literally the exact same ones our ancestors ate to match our nutritional needs. Humans didn't eat seaweed or algae or the deepest sea fish during most of our evolutionary history, but now much can be harvested from the oceans—which are less depleted than our forests. Genetic engineering may also help us if we're careful with it. I'm not in agreement with foods touting "non-genetically engineered" as making a refined product of low nutrition desirable. We can potentially put back in vitamin and antioxidant levels that have been slowly bred out of our plants over all the years when taste or appearance were being emphasized.

Proposals for regulating our diet may sound extreme, but I can remember the days when cigarette companies had people convinced that similar statements about cigarettes were radical and were still describing well-established adverse health effects of tobacco as "unproven." There are many parallels between smoking and obesity. The U.S., which once led the world in rates of smoking and lung cancer deaths, is now setting an example for smoking cessation, antitobacco legislation, and declines in lung cancer. We're again at the front of the obesity epidemic and it's time to start turning this around too.

6

Defending Home, Hearth, and Hedge Fund

"Merlyn," said the king, "tell me the reason for your visit. Talk. Say you have come to save us from this war."

—T. H. White, *The Book of Merlyn*

When T. H. White began writing the concluding volume of his wildly popular *Once and Future King*[1] series, in which he recounted the adventures of King Arthur and the knights of his Round Table, he intended for "that one brief shining moment that was known as Camelot" to represent something quite different from what emerged in the published version or the glitzy musical and film it spawned. Merlyn was indeed appearing before Arthur on the eve of impending war with his son Mordred, trying to "save them." "I have suddenly discovered," White wrote to his former Cambridge tutor, "that the central theme [of the saga of Arthur] is to find an antidote to war."[2]

White's main point—that violence always begets violence and war is never justified—is framed in distinctly evolutionary terms and uses examples from animal ethology. Merlyn delivers an antiwar—and somewhat antimankind—lecture and then demonstrates alternatives with King Arthur's animal mentors from the earlier books. Merlyn tells Arthur the following (playing perhaps a bit loose with the facts):

- Between 1100 and 1900, the English were at war for 419 years—the French for 373.
- Nineteen million men were killed in Europe every century.
- The amount of blood spilled in European wars alone would feed a fountain running 700 liters an hour since the beginning of history.
- Most exceptional men from Socrates on who've tried to give the bloody hordes advice against war have been murdered for their trouble: in fact, Europe's favorite religious myth is of God becoming a man, coming down to earth to deliver truth—and being murdered.

The animals chime in to observe that "the so-called primitive races who worshipped animals as gods were not so daft as people choose to believe." They ridicule man for naming himself "Homo sapiens," seeing little evidence of wisdom and suggest "Homo ferox" instead: the most ferocious of all animals. They point out that

- Man is unique as an animal that kills for pleasure.
- Man even trains other animals to kill for his pleasure.
- Man is the only animal that slaughters his own kind en masse.

Some animals interject that actually ants are the one other species to behave as badly with their own. In a dream scene, Arthur finds himself briefly an ant. Two ant colonies' simple-minded warriors repeat platitudes as apt in our times as in White's (try substituting "oil" for "syrup"):

- We are more numerous than they are, therefore we have a right to their syrup.
- They are more numerous than we are and are wickedly trying to steal our syrup.
- We must attack them today or they will attack us tomorrow.
- We are not attacking them at all but offering them incalculable benefits by taking over with our superior control.

The dream's next sequence is a mock court where animal lawyers argue the cases for and against war. The prowar case is that war is the only method by which humans will practice birth control and that, as a satisfying outlet for violence, it reduces the drive for child abuse, soccer stadium riots, and overzealous dentistry. Its strongest "plus" is the slight chance that it might eliminate man entirely. Lest you think that White is entirely facetious, as he wrote this he lived alone on the coast of Scotland, with no friends save a few correspondents. His sole companions were an owl he'd nursed back to health following a wing injury and two pet geese. He was more aloof from human attachments than Tinbergen and as fond of animals as Lorenz.

The arguments against war included the suggestion that people might better be regularly injected with adrenalin—as that seems to be the point of it all. Instead of the present conventions, all officers on the losing side should be executed at its end—irrespective of specific "war crimes," the animals said—so that the outcome would be as suspenseful for those declaring war as for those sent to fight it.

Arthur awakened and offered Mordred half his kingdom. "Mordred accepted but Arthur had been prepared to give him all if he'd refused."

The previous *Once and Future King* books were runaway best sellers—the most popular children's books of their time and crossing over into hearty adult sales, much like Harry Potter for this generation. So did White's publisher race *The Book of Merlyn* to press? Hold midnight release parties?

No.

The Book of Merlyn was not published during White's lifetime. It was 1940 and England was embroiled in World War II. Pacifist messages are invariably suppressed in wartime—ironically the only time they really matter. White's publishers incorporated just enough scenes from *Merlyn* into the fourth volume to make the series end coherently. Their version left the impression that "chivalry" was the ultimate ideal. The Round Table's absolute principles and laws could be enforced by execution, ignoring extenuating human circumstances.

Their Arthur, after trying to reason with Mordred, ended up reluctantly, but righteously, at war.

It was, in fact, 1977—peacetime following an unpopular war—when *Merlyn* was finally published by a small academic press in the United States. Reception was positive if uneventful and trade houses in both the U.S. and England put out their own editions. This passionately pacifist—albeit chillingly cynical—little volume achieves what William James meant by "making the ordinary look strange." It poetically touches on many of the ethological points about human violence that we can examine from a more detailed scientific perspective.

The Killer Ape

Man biologically considered . . . is simply the most formidable of all the beasts of prey, and, indeed, the only one that preys systematically on his own species.

—William James

For all the clichés about animal aggression and living in a "dog-eat-dog" world, man is indeed the most violent animal with his own species. In the tradition of James's "most formidable predator" and White's "homo ferox," anthropologist Raymond Dart coined the term "Killer Ape" for the hypothesis that man ascended to domination over other species because of his unprecedented aggression. Actually, it seems more like the level of our present violence resulted from our population density and technology rather than leading to it, but the designation Killer Ape remains apt. Before directly examining aggression, and especially the mass aggression of war, let's consider three related concepts: territoriality, private property, and the perception of threats. As White noted, these three things account for much of the motivation of modern war.

We think of our homes as sacred. We lock them. If someone tries to invade—even our yard in some communities—we call the police. Permanence is valued, as in "My family has lived on this land for generations." Having to relocate because a highway is being built or one can no longer care for a large house feels *wrong*, a major loss.

Modern life is criticized for uprooting people—as if multiple moves were abnormal. We see "street people" who sleep a few nights at a time in one shelter or rotate between several spots in parks and doorways as pitiful. At best, a nomadic existence is fitting for the young or immature—we expect that artists or students might move frequently for a few years before settling down.

But just what are instinctive versus abnormal patterns for humans? In Tinbergen and Lorenz's ethology, the term "territory" refers to any geographical area that an animal consistently defends against others of its species and, occasionally, animals of other species. Robert Ardrey popularized the idea that this was an important factor in human as well as animal psychology in his book *The Territorial Imperative.*[3] Through the 1960s and '70s, social ethology placed a huge emphasis on the importance of territory. Concepts of personal property, ownership of land, and violent defense of these were explained as arising from this "imperative" or instinct.

In fact, only a minority of species maintain territories with well-defined boundaries that they defend. These species tend to be our distant relatives, including various fish and birds. Territoriality is also rarely a fixed trait of a species. As described in Chapter 2, Tinbergen discovered that male snow buntings defended territory when courting mates and that both genders did so when feeding their young, but neither acted territorially at other times of the year. Male yellow warblers defend their territory only against other males, and females against females. Nectar-feeding birds defend territories only in the morning, when plants are richest in nectar. Grizzly bears chase away intruding bears when food is scarce. But during salmon migration, they cheerfully crowd in next to each other along riverbanks like happy hour patrons at a bar.

Many more species have an area that an individual or group frequents but does not defend; biologists call this a home range. The home ranges of different groups overlap; in the overlap areas, the groups avoid being in exactly the same spot rather than trying to expel each other. Other species habitually wander after food or other resources without even a home range.

Humans, by nature, fall into this last category. As I quoted evolutionary psychologists Leda Cosmides and John Tooby in Chapter 2: "Each of our ancestors was, in effect, on a camping trip that lasted an entire lifetime. . . ."[4] Ten thousand years ago, before agriculture, humans were entirely nomadic. They followed game, changing rainfall or temperatures, many roaming hundreds of miles in a lifetime. The world's population numbered 5 million people scattered around the globe's habitable land. They roamed in groups as small as a nuclear family or as large as several dozen, with an average density of about one person per square mile. Even now that the remaining hunter-gatherer tribes are severely curtailed as to where they can go because of agricultural land claims, they observe at most a "home range" pattern, relocating camps over areas of many miles.[5]

So where do our instincts to guard our homes and yards—and even national boundaries—come from? Most likely from instincts about protecting our nightly campsites. In other nomadic species, when two groups converge on a scarce resource such as a watering hole, one group waits patiently until the other leaves. If avoidance is not perfectly honored, feathers ruffle, scaly spines raise, or teeth bare, and etiquette is restored without an actual fight. When humans were spread one per square mile, they easily avoided each other with only occasional threat displays. With the advent of agriculture and the impracticality of digging up and replanting crops, we began to stay in one place. But even then, people's land was shared by a group and had vague boundaries. They weighed the cost of defending it before objecting to intrusions.

Gradually, growing population density made for sharper boundaries and greater vehemence about keeping others out. I can remember being taught American history in the 1960s by teachers who ridiculed the Native Americans' inability to understand the fundamental principal of private property as they signed European documents. Now, politically correct thinking emphasizes how they were defrauded, "robbed" of "their land." But we still don't take seriously what the natives were saying: *No one owns the land.*

Today we live in cities with densities as great as 11,500 people per square mile[6] and are constantly confronted by encroaching strangers. We use a number of strategies to tolerate this (see Figure 1) but accidentally bumping against one another or taking parking spaces someone else believes he spotted first still account for numerous urban fights, sometimes fatal. White mentioned soccer audience violence, but the record-breaking riot was not to occur for another six decades.[7]

Similarly, humans have always had instincts about personal possessions. But they protected a bit of extra food they'd gathered for the next day, a spear they'd chiseled, or a bit of hide clothing. "You can't take it with you!" had a more earthly meaning back then. It didn't make sense to own more than you could carry to the next campsite. Once crops tied us to a place, we began to erect stone and brick buildings. These served many purposes including better shelter from the elements, but keeping others out was always a major factor. You didn't need to take it with you. People could store a vast supply of the new dryable grains, tools, and clothing. The most powerful individuals and families began to stake claims to increasingly large stores of these necessities . . . and to luxury items.

Humans have always traded: meat for gathered vegetables, a carved spear for a clay pot, even red ochre body paint for shells strung onto a necklace. These have all been documented among hunter-gatherers 100,000 years back. But once the ability to store goods in permanent fortresses took hold, the demand to trade longer term commodities grew—excess grains grown this season to a farmer with failed crops in exchange for double that amount next year or for work to be done for the grain holder. As society invented more abstract ways to represent food, land, and labor with money and credit logs, one individual could amass personal property worth a hundred or thousand times that of another.

There is also, obviously, a compelling instinct to provide for one's offspring; this is practically synonymous with whose

FIGURE 1A

Interpersonal Space in Elevators

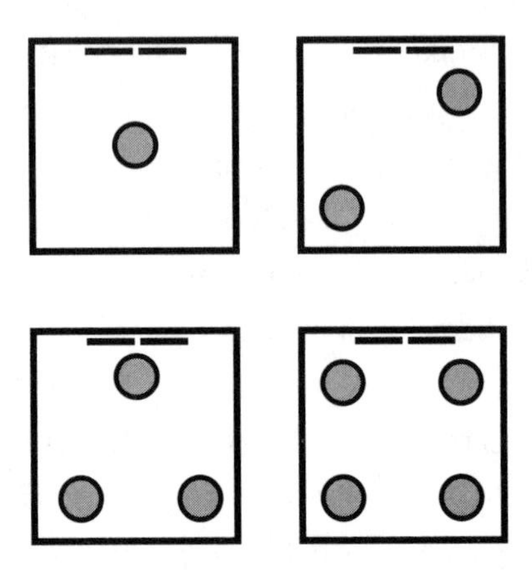

Instinct determines what we consider our personal space in public places. Social psychologists find that individuals position themselves by rigid, if unconscious, rules as they ride elevators or sit at public tables. *Above*: One person stands near the center or near the control panel. When a second gets on, the two position themselves diagonally for maximal distance. With three, two are in the back corners and the third stands either in front of the doors or in the front corner by the control panel. Four occupy each corner and a fifth would stand in the center, etc. Couples, families, or friends getting on together occupy one position.

Above opposite: Rules at library or cafeteria tables are similar. If one expects strangers may arrive and share the table, the first person instinctively sits in a corner indicated by the

genes will survive. However, people previously provided for their offspring mostly until maturity, with occasional provisioning for them and for grandchildren if the family remained together. Now the powerful and rich can direct these instincts at supernormal family estates, trust funds that endure for generations, and, in the case of monarchies, permanent rulership for the family. (If recent American politics is any guide, this may even apply within a democracy.)

We decry vicious land feuds but we don't stop to ques-

FIGURE 1B

Interpersonal Space in Public seating

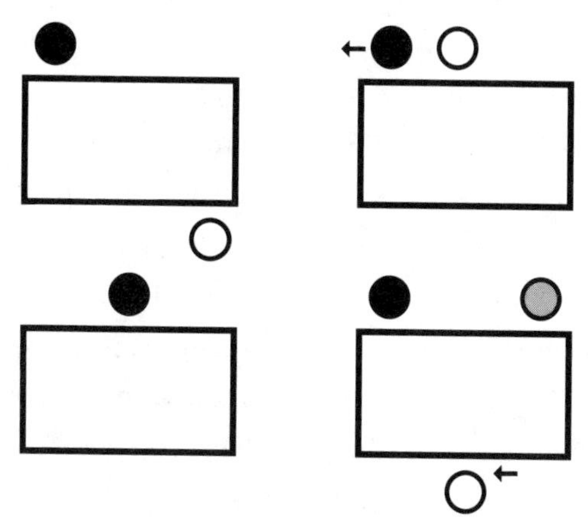

black circle (*top left*) and the second arrival knows to sit in the place of the white circle. If a social psychology researcher arrives second and sits down in the position indicated by the white circle (*top right*), the intruded-on person makes defensive and annoyed body gestures, and often departs for another table. Assertive individuals trying to keep their table for themselves position themselves like the black circle in the lower left diagram. Any new arrival will choose to join a table with the first seating pattern until those are taken. If someone does eventually join the middle-seated person, the first individual usually shifts position to establish the same two-person seating pattern as in the upper right. A third will generally cause the group to rearrange themselves (*bottom right*).

tion the unnaturalness of the whole premise. Jurisdictions from Israel to the southern United States have home protection laws that allow people to kill anyone even trying to break into their home.[8] More liberal countries may declare that the poor should have the same access to schooling or health care as that of the rich, but one rarely hears the suggestion that they are entitled to equally good land. Everywhere, the rich wall off the best beaches, mountainsides, and even private islands for their exclusive use.

Karl Marx, in advocating the abolition of Capitalism, paid some attention to its evolutionary history. Hunter-gatherer societies, he wrote, were structured so that labor and its economic reward were virtually identical. "Pastoral" societies—that is, agricultural ones—developed landowners and rulers who exploited serfs and siphoned off a proportion of the results of their labor. However, economics was still directly tied to the product of labor. With Capitalism, labor and its product are separated, alienating the system from nature. I believe that a crucial reason Marxism has failed to bring about more improvement in society is that Marx didn't take into account many other lessons from evolution, ethology, and sociobiology. The effects of crowding, the unnatural settling of a nomadic species, and the supernormal stimulation of instincts for acquisition all militate against a redistribution of resources.

Threat Perception

T. H. White also recognized how much violence we commit in the name of protecting ourselves. Countries have "Departments of Defense" not "Departments of Aggression." Whether we're hearing an account of a schoolyard fight or a war between nations, the narrator is never the aggressor, always the defender. Israeli tanks bulldoze homes to retaliate for suicide bombings; Israeli soldiers shoot boys throwing stones to protect themselves. Palestinians strap bombs on themselves and blow up civilians in cafés to retaliate for the stealing of homes, the shooting of youths, the assassination or jailing of a leader. Irish Catholics retaliated against the

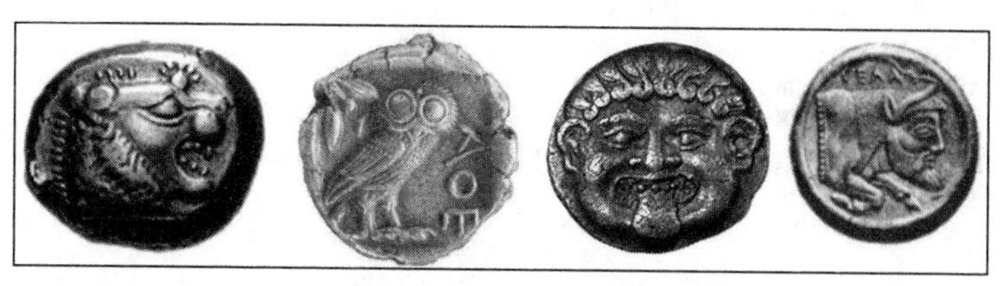

The Lydian Lion (*left*), made of a gold and silver alloy, was probably the world's first coin. It was minted in Sardis, Lydia, Asia Minor (present-day Turkey) circa 610 BC. Greek and Roman coins that depicted animals (owl drachme, *middle left*), rulers, deities, mythological creatures (Medusa, *middle right*), or some combination like the Sicilian river-god Gela (man-headed bull, *right*) soon followed. These precious metal coins and decorative shells were strung onto trading necklaces and served as a transition between objects that had inherent value being traded and a purely abstract currency.

"Cash, Jordan. That's what separates man from the apes."

Protestants who retaliated against them. We must do this to stop the violence, they say. Then the other side does something violent—also to stop the violence. It's easy to find other recent examples: Turkish and Greek Cypriots post pictures of dead and injured children on billboards at the border, telling tourists what atrocities the other side committed against them. Shiites seek revenge on Sunnis for the revenge they sought on Shiites.

Many of these ongoing feuds involve what sociologists call a "double minority" situation. When one is outnumbered by the enemy, one feels more threatened and more justified by using any means of aggression. In the above conflicts, Palestinians are a minority within Israeli borders, while Jews are a minority in the larger region. Protestants are a minority in Ireland, but Catholics within Great Britain. Sunni are a minority within Iraq but Shiites within the whole Arab world. This can lead to each side defining itself as the underdog for whom any guerrilla tactics are justified against the stronger enemy.

We have instincts for sensing threats, but like our other instincts, they are honed for risks encountered wandering the savannah. The amount of fear generated by a potential threat versus its likelihood of harming us are not tightly related in the modern world. Evolutionary psychologists pointed this out in explaining why people became so terrified of anthrax. In 2001, anthrax killed five Americans. In other recent years, zero. In 2001, in the United States, 36,000 people died from influenza. Another 11,000 people died from respiratory syncytial virus, or RSV.[9] *Have you ever even heard of this infection?* Anthrax was novel, involved a bad person, and it was not by random chance that people contracted it. Ditto terrorism. The 3000 people who died in the World Trade Center are hardly a minor number. But 41,730 died in traffic accidents in 2001[10] without generating the same terror. The following year, an editorial in an Israeli newspaper pointed out that significantly fewer Israelis had died from terrorism than from traffic fatalities, and an outcry arose against these "insensitive" remarks. But what we really are insensitive to are the highway deaths. Surely families of those lost to suicide bombings or 9/11 and families of people dying in fiery car crashes experience equivalent anguish. What's not equivalent is the anticipatory terror for self and loved ones that is stirred up in people who are strangers to the deceased. Our defensive instincts aren't triggered strongly by traffic accidents. We're much more vigilant against villains.

In the brilliantly titled, "If Only Gay Sex Caused Global Warming," Harvard psychologist Daniel Gilbert wrote, "No one seems to care about the upcoming attack on the World Trade Center site. Why? Because it won't involve villains with box cutters. Instead, it will involve melting ice sheets that swell the oceans and turn that particular block of lower Manhattan into an aquarium."[11] He outlined four features of overrated threats:

1. They are the product of human intention.
2. They violate our moral sensibilities.
3. They represent an immediate problem.
4. They appear suddenly or grow rapidly.

We perceive the malevolent-human type of threat too easily in our overcrowded world and our instincts for what to do when we really encounter it are also off.

An Eye for an Eye

Humans recognize the principle of retaliation. When children fight, parents investigate who hit whom first. We often think it's okay to "hit back" but never to "start a fight." Most cultures teach "self-defense," not "aggression." But we have a sense of proportion—you can hit back, but not too hard. An eye for an eye, but not for an eyelash. When countries are threatened, whether by outsiders or their own rebels, they are expected to defend themselves, but onlookers will decry a "disproportionate" response.

However, we judge proportion no better than we judge threats. Research suggests that people actually *try* to retaliate proportionally. In a study conducted at University College London,[12] pairs of volunteers were hooked up to a device that allowed each of them to exert pressure on the other's finger. The researcher began by exerting a fixed amount of pressure on the first volunteer's finger. The first volunteer was then asked to exert exactly the same amount of pressure on the second volunteer. The second volunteer was asked to exert the same amount of pressure on the first. The two took turns applying "equal" amounts of pressure while the researchers measured the actual amount of pressure they applied.

Each volunteer believed he or she was obeying instructions and responding to the other's touches with equal force, but they typically responded with about 40 percent more pressure than they'd just experienced. What began as a series of soft touches quickly became moderate pokes and then hard prods. Each volunteer believed that he responded with equal force and that, for some reason, the other was escalating. Neither realized that the escalation was the natural by-product of a neurological principle that we notice a touch we receive much more than one we deliver.

Other research shows that, even when physical blows are not involved, we think of the causes and consequences of our own actions very differently from how we view the causes and effects of anyone else's. In a University of Texas study,[13] pairs of volunteers played the roles of world leaders deciding whether to initiate a nuclear strike. The first volunteer was asked to make an opening statement, the second volunteer to respond, the first to then respond to the second, and so on. At the end of the conversation, the volunteers were shown several of the statements that had been made and asked to recall what had been said just before and after each of them.

The results revealed a dramatic asymmetry: When volunteers were shown their own statements, they clearly remembered what had led them to state this but were vaguer on what had resulted. When they were shown the partner's statements, they remembered much better how they had responded than what had gone before. In other words, volunteers remembered the causes of their own statements and the consequences of their partner's.

It's obvious when we think about this: because our senses point outward, we can observe other people's actions but are less aware of our own. As mental life is an internal affair, we can observe our own thoughts but not the thoughts of others. Our reasons for our own punching will always be more salient to us than the punches themselves—but the opposite will be true of how we perceive other people's reasons and other people's punches.

Surrender

Virtually all animals have surrender signals that reliably end within-species battles before they fight to the death. Wolves bare their throat to the victor; gorillas just lower their head before another's dominant stare to stop a confrontation. One way that human mercy has gone awry is that the long-distance technology of modern warfare prevents aggressors from seeing appeasement signals before striking. Another is that they no longer see the graphic results of their aggression. This applies not only to our modern technology of far-flung missiles, but also to early postagricultural warfare. The Crusades was one of bloodiest conflicts ever, but it was a "modern" war by anthropologists' standards: fought with shields and helmets, riders on horseback, and battering rams or catapults for assaulting walled cities. No hunter-gatherer tribes struggling over scarce resources were ever this removed from those they fought.

Modern fights do, of course, employ surrender signals when they are just one-on-one street fights. One could theoretically beat a weaker opponent to death. But people give up, apologize, beg. Or the winner sees the damage he has inflicted and doesn't wish to go further. That's why very few fistfights end in death. More knife fights can—they are swifter, with less chance to surrender or see gradual injuries. This is also why homicides correlate so strongly with gun ownership—shooting is yet more removed. In most unarmed confrontations, no blows are even exchanged. Opponents posture, puff up their chests, threaten. They measure their opponent—one backs off. War doesn't allow for this.

Pseudospecies

Of course not all war is explained by distance. Killing happens up close, too. Human-on-human violence is as old as mankind. As T. H. White, William James, and numerous others noted, it is more extreme than for most species. Not only *The Killer Ape* but also *The Demonic Ape* and *The Most Dangerous Animal* all refer to man. Why is this? Erik Erikson, a psychoanalyst who met Konrad Lorenz in the years after World War II, proposed the concept of the "pseudospecies":

> The term denotes that while man is obviously one species, he appears and continues on the scene split up into groups: from tribes to nations, from castes to classes, from religions to ideologies . . . which provide their members with a firm sense of unique and superior human identity.[14]

Our "pseudospecies" are those who look like us, believe the same things, or speak our language. Erikson said the pseudospecies provides people with a positive sense of identity but also obliterates our sense of other humans as our kin. Any national or religious identity always involves a myth of being the superior or chosen ones. Its dark side is a projection of negative, inferior, or evil traits onto other groups.

Culture and the tutelage of prolonged childhood allowed each pseudospecies to inculcate its glorified identity and negative prejudice against others. After this schooling, man applies his instinctive rules about how he should behave toward his own species only to his pseudospecies and "possessed by this combination of lethal weaponry, moral hypocrisy, and identity panic is apt to . . . turn on another subgroup with a ferocity generally alien to the 'social' animal world," Erikson said.[15]

Great religious leaders have always taught that mankind is one great whole but, as Erikson pointed out, their churches "have tended to join rather than shun the development which we have in mind here, namely, man's deep-seated conviction that some providence has made his tribe or race or caste, and, yes, even his religion 'naturally' superior to others."[16]

Political leaders play to this instinct because whipping up paranoia about enemies consolidates their power. Leaders on two sides of a conflict take turns demonizing each other . . . literally. Iran's Khomeini routinely referred to the United States as "the Great Satan." Bush included that country in his "axis of evil." Most colorfully, Venezuelan President Hugo Chavez told the United Nations, "The devil came here yesterday." Speaking at the same lectern used by Bush the previous afternoon, Chavez crossed himself, "and it smells of sulfur still today." Each leader knows the game furthers both careers. War or even threat strengthens national pseudospecies' identity and mobilizes instincts to follow a leader in time of crisis with less evaluation than when there's time to think. It makes people cheerfully forego personal goals on which they would insist in calmer times. Defeating the enemy becomes paramount. Protest is squelched. People who understand other cultures, speak other languages, or appreciate diversity are ignored or even ostracized.

In a case of life-imitates-art-imitating life, the 1997 film *Wag the Dog* featured an American president hiring a Hollywood producer to trump up a fake war with Albania. Commentators drew parallels to George W. Bush and events of the second Gulf War, such as the manipulation of media around the Jessica Lynch story because the invasion of Iraq occurred shortly after the film's release. However, *Wag the Dog* was based on a 1993 novel that (unlike the film) identified the president as George *H.* Bush and the war as the *first* Gulf War, while satirically suggesting Bush had hoped to win reelection by emulating Margaret Thatcher's invasion of the Falkland Islands.[17]

America with its melting pot, land-of-opportunity attitudes may have unusually literal boundaries for its definition of pseudospecies. Though some fundamentalist religious or ethnic subgroups within the country stick to more traditionally small pseudospecies, one expressed to me by a newly arrived

"It really shook me, I can tell you. I dreamed the meek inherited the earth!"

Morrocan cabbie seems apt. He observed that very few countries in the world are as generous as America in feeding, clothing, and educating strangers or granting them equal rights and opportunities once they are within their borders, but few are as callous about taking human life if it lies outside their borders.

The military of each country promote even cruder propaganda about the enemy than do politicians. The other side is portrayed as taking pleasure in killing and committing atrocities, while our own side is only responding responsibly and doing what is necessary—"fighting to end war." Every war has its epithets—"infidels," "gooks," "towelheads"—to promote an image of the enemy as less than human. "The Red Menance," "the Yellow Peril" suggest bacterial diseases to be eradicated. Before and during the Rwandan genocide, Hutu media consistently referred to Tutsi as "cockroaches."

The Adrenaline Rush of War

Foreign correspondent Chris Hedges's book *War Is a Force That Gives Us Meaning*, described what most veterans but few others know—war's dark appeal—its sense of intensity, nobility, elevation above the vapidness of everyday existence. Especially in bad times, war imparts purpose and hope, unites neighbors in a common cause. Hedges used few biological terms except the analogy to drugs: "The rush of battle is a potent and often lethal addiction. . . . The chance to exist for an intense and overpowering moment." However, it's easy to translate his words into the adrenaline associated with a fight which overrides fear, pain, and altruistic considerations.

Hedges intuitively grasps the concept of pseudospecies. "Most national myths, at their core, are racist," he says, "and even as war gives meaning to sterile lives, it promotes racists and killers."[18] Each chapter of his book begins with a quotation about how leaders are whipping up the masses with disinformation or how each side sees itself as the innocent victim of the other. He sets us up to hear these as referring to Saddam Hussein, Al-Qaeda, or Bush . . . and then reveals the timely observations to address the Trojan War, Oliver Cromwell, or World War I.

Hedges says that nation, God, and nobility of cause unite the nation at home behind an offense-in-defense's clothing. These notions also draw recruits into the military. But nation, God, and nobility don't keep soldiers fighting. Hedges describes what I've heard from veterans in my practice: that most soon see the horrific, ignoble underside of war, the pointlessness of the cause; they develop a cynical perspective on leaders. Soon soldiers are not fighting for these things, but for their lives. Even if there was no real threat at home, there certainly is once one goes into a war zone—or starts a war in a zone that wasn't at war. The paranoid belief is self-fulfilling and it becomes kill or be killed.

The ties to the broader group back home—by whom the soldier often feels forgotten, tricked, or betrayed—weaken. The primary bond is now with one's immediate peers—the other soldiers at one's side. They save each other's lives repeatedly, even as they lose other buddies. Most veterans remember combat as the most intense period of their lives, and the bonds formed there as the closest. The bonds within the nation back home only dimly echo this. Ironically, at this stage, the momentum of war is furthered by human's altruistic, cooperative instincts. Many animals would defend only a direct attack on themselves, but troops are defending their buddies or warding off a potential attack on the next platoon.

Hedges describes the atrocities that eventually drove him from war reporting—babies impaled on bayonets, heads stuck on fenceposts, mutilated body parts fashioned into coin purses and other war mementoes. The extreme acts that humans commit under the full sway of the inhuman pseudospecies functioning are virtually the same in what he recounts from Bosnia and Iraq and in what I heard from Vietnam Vets. But Hedges adds, "Like every recovering addict there is a part of me that remains nostalgic for war's simplicity and high, even as I cope with the scars it has left behind, mourn the deaths of those I worked with, and struggle with the bestiality I would have been better off not witnessing."[19]

Can We Ever Stop War?

"Only the dead have seen the end of war," said Plato. Is this true? Short of genetic manipulation, I think we shall never see the end of human violence or murder—though genetic manipulation is not impossible in our future (I address it in the last chapter). But, assuming human instincts remain the same, is it possible to roll back war, to reduce it to the fights—only very rarely to the death—of pre-agricultural times?

The technology for war is never going to disappear. Once gunpowder, atomic fusion, or weaponized anthrax have been invented, they cannot be uninvented. But the technology of propaganda—the mass communication that can whip up violence by reinforcing boundaries between pseudospecies and create supernormal versions of threatening enemies—has just as much potential for communicating similarities between groups, for showing us surrender signals and the suffering of our enemies up close. As Hedges says in explaining why war propaganda exists, "A soldier who recognizes the humanity of the enemy makes a troubled and ineffective killer."[20]

The Internet especially—because it's user controlled—but also satellite television, text messaging, and all other new media have potential to help a person on one side of the world bond with someone on the other. So far, the vast potential of the Web is not much used this way. Americans look mainly at Western sites geared to their own point of view, fundamentalist Moslems at ones geared to them. But it is at least more possible to override the information provided by one country's leader and its own press. In the run-up to the Iraq war, Americans could read European newspapers and learn what the American media said only long after the invasion: there was no evidence of "weapons of mass destruction" under development in Iraq. We can read the English Al Jezeera or Al Ahram sites if we want to know what Arabs are thinking. Iraqis could read anyone else's news to see that they weren't repelling the invasion no matter what their Press Minister claimed. Not many people did in these instances, but antiwar movements may yet learn to make better use of the Web and other media. It has vast potential to reverse the overstimulation of instincts for defensiveness and aggression and begin stimulating urges of peace and reconciliation.

It also may be time to intellectually override our instincts about who should lead. It's no coincidence that the vast preponderance of the world's leaders are men. Though some would say this is because of social conditioning about males-as-leaders, evolutionary psychologists see a biological instinct at work. In paleolithic times, men did lead certain activities—hunts, fights—and there weren't senates or courts for *anyone* to lead. Thus men inherited stronger biological underpinnings that lead them to seek office and make them likelier to win elections. This is not at all equivalent to doing a better job once in office in our crowded world loaded with technological weapons. Women through history have said—as do contemporary ones in Israel and Palestine—that a group of mothers could sit down and hash out in one afternoon an agreement that has eluded male rulers for years. Extrapolating from both animal research on testerone's effects and simple statistics on rates of human violence, women might be better at some of the tasks facing us today—ending wars, helping the environment. Or a half-male, half-female government—where those people are chosen from among the most cooperative rather than the most competitive—might lead more wisely.

"Democracy is a device that ensures we shall be governed no better than we deserve," George Bernard Shaw observed. We teach that democracy is the ideal form of government as if it ensured human decency. We tend to forget that many horrific leaders and regimes were elected, including Hitler and the Nazis. As James Bovard advised, "Democracy must be something more than two wolves and a sheep voting on what to have for dinner."[21] Early in America's history, the Bill of Rights was added to the Constitution to put checks on the wolf-majority effect. But the Bill of Rights applies only to U.S. citizens and domestic issues. We need similar restrictions on what we can do to other groups outside our borders.

It sounds crazy to Americans, and most nations, but Costa

Rica has disbanded its army and directed that money into environmental concerns. Far from being overrun by aggressors this stance has afforded it peace with its neighbors. Most Americans cannot picture their disarming without being overrun by one or another bugaboo: Communists, Jihadists. But if we imagine the Soviet Union of the 1970s and '80s had ceased pouring all assets into its monster military and instead directed that capital into education and technological development, we hardly see it having been invaded by the U.S. or western Europe. Rather we imagine it might have saved itself from downfall. Countries, including ours, need to seriously question whether displays of force really diminish the number and effectiveness of their enemies or in fact raise the danger as supernormal stimuli to the defensive instincts of others. I'm not trying to claim disarmament is obvious or simple, just that it should be one of the options on the table as we debate national security or negotiate among nations.

7

Vicarious Social Settings from Shakespeare to *Survivor*

Television is the first truly democratic culture—the first culture available to everybody and entirely governed by what the people want. The most terrifying thing is what people do want.

—Clive Barnes

"Chewing gum for the eyes" is how Frank Lloyd Wright described television. With its canned laugh tracks and illusory "friends" available at the flick of a switch, it occupies a huge place in Americans' lives. The average household has 2.7 people and 2.9 television sets. Americans watch almost five hours a day of television—the majority of their leisure time and more than any other single activity besides work or sleep. Children and adolescents watch more yet.[1]

"Boob tube," "idiot box," "one-eyed monster." The nicknames betray our concern even as we try to laugh it off. Just as people cheerfully admit to being couch potatoes who eat junk food, self-proclaimed TV addicts who are glued to the tube rarely try to wean themselves from the "glass tit." Fifteen years ago, researchers posed a question to children of intact families: if they had to give up either their father or their television for one week, which would go? A *very* slight majority deemed their dads more essential; who knows if they'd win today.

How has television become so central to people's lives? There's much to decry about its content—violence, sexism, and relentless advertising—but the most sinister aspect of television lies in the medium itself. There's a growing body of research on what it does to our brain; "idiot box" is not far off.

Humans have a basic instinct to pay attention to any sudden or novel stimulus such as a movement or sound. In 1927, the legendary Russian neurologist Ivan Pavlov named this reflex the "orienting response." Shared with other animals, the orienting response is part of our evolutionary heritage. It evolved to help us spot and assess potential predators, prey, enemies, and mates. The orienting person or animal turns eyes and ears in the direction of the stimulus and then freezes while parts of the brain associated with new learning become more active. Blood vessels to the brain dilate, those to muscles constrict, the heart slows, and alpha waves are blocked for a few seconds.

By the age of six months, babies orient when a television is turned on. Adults continue to do so. Even researchers studying the effect are not immune: "Among life's more embarrassing moments have been countless occasions when I am engaged in conversation in a room while a TV set is on, and I cannot for the life of me stop from periodically glancing over to the screen," confesses Percy Tannenbaum of UC Berkeley. "This occurs not only during dull conversations but during reasonably interesting ones just as well."[2]

The visual techniques of television—cuts, zooms, pans, switches from one camera angle to another within the same visual scene, and sudden noises—all activate the orienting response.[3] The effect persists for four to six seconds after each stimulus. Producers of educational television for children have found that judicious orchestration of these formal features can increase learning—presumably by keeping children focused on the screen. After a certain level of intensity, however, the orienting response is overworked and effects on learning and attention begin to reverse. This is what we see with ads, action sequences, and music videos, where formal features provoke orienting at the rapid-fire rate of one per second. Following prolonged bombardment with these stimuli, the viewer develops a strange mix of physiological signs of high and low attention. Eyes stay focused, the body is still and directed toward the set, but learning and memory drop to lower levels than when not orienting. Measurements of metabolism, including calorie-burning, average 14.5 percent *lower* when watching TV than when simply lying in bed.[4] EEG studies similarly find less mental stimulation, as measured by alpha brain-waves, during viewing than during reading or other quiet activities.

When researchers query people watching television, most viewers report feeling relaxed but passive and not alert—in keeping with their physiological signs. The minute the set is turned off, the relaxation ends, but the passivity and lowered alertness continue. Many feel as if television has somehow sucked out all their energy. People say they have more trouble concentrating after viewing than before. After watching TV, most people's moods are either about the same or *worse than before.*[5] In contrast, few indicate such problems after reading. After playing sports or engaging in hobbies, people report improvements in mood.

Lack of energy and ability to focus as well as poorer moods are short-term negative effects. What about the long term? An article in *Pediatrics* reported that the longer children sit in front of a TV, the less likely they are to sleep well.[6] Children who view more television have more behavioral problems including attention deficit disorder and, on average, perform worse academically. One study of 2000 children found that watching television before the age of three was linked to poorer reading and math skills at the ages of six and seven. For children this young who watched more than three hours of television per day, the negative effect on skills was similar to that of having a mother with a very low IQ or little education.[7] Psychologists have suggested that the cause of this may be that the insistent noise of television interferes with the development of "inner speech" by which a child learns to think through problems and plans.

Adults who watch more television also report a higher rate of mood and intellectual difficulties. The TV industry argues that this doesn't prove causation: people with these problems might then be inclined toward excessive viewing. However, one study, now two decades old, seems to demonstrate direct causation between television and many of these difficulties. In the 1980s, researchers at the University of British Columbia got a unique opportunity when several remote Canadian mountain communities, which had never before had broadcast television, contracted to receive it by cable. The researchers surveyed people's activities and tested them on a variety of cognitive and behavioral measures before television was installed and for five years after. *Immediately*, the number of hours devoted to sports declined along with dancing and all other physically active leisure pursuits. Over time, both adults and children in the town became less creative on problem-solving tests, less able to persevere at tasks, and less tolerant of unstructured time.

In the article, "Television Addiction," psychologists Robert Kubey and Mihaly Csikszentmihaly observe, "Perhaps the most ironic aspect of the struggle for survival is how easily organisms can be harmed by that which they desire. The trout is caught by the fisherman's lure, the mouse by cheese. But at least those creatures have the excuse that bait and cheese look like sustenance. Humans seldom have that consolation."[8]

Animals and man are indeed often harmed by what they desire—especially when encountering new stimuli for which evolution hasn't prepared them. That's the central thesis of this book. But I believe that the human television viewer has very much the same "consolation," "excuse," or explanation as the fish on the hook—the supernormal stimulus. The trout lure mimics—or exceeds—the darting and bright coloration of the tastiest fly; television mimics—or exceeds—the adventures, athletics, and human connections that we desire but actually miss out on as we stare at a box of electric circuitry.

Scheherazade's Game

> *Addictive culture deals with issues people feel to be crucial in their lives, but instead of confronting these issues, addictive art merely re-confirms the values and internalized pressures that produced the issues in the first place. Addictive art is briefly palliating; the relief lasts only as long as the art does and one is left needier than before.*
>
> —Anthony J. Cascardi, *The Cambridge Companion to Cervantes*

Star Wars is the most popular film of all time. But after viewing it, did audiences come away satisfied and happy to reflect on what they'd experienced? No. *Star Wars* generated one primary impulse—the desire for a sequel—for which purpose large pieces of the plot were left hanging in midair at the end of the film. Indeed, even as newspapers reported the record-breaking box office grosses, a screenwriter was working on a script for the first sequel, which already had its production dates locked into the studio schedule.

We may think of "Blockbuster Movie VI: The Endless Sequel" as a modern phenomenon, but literature has a long history of hungrily awaited installments in neverending series. Nineteenth-century audiences waited for the next Charles Dickens novel or Sherlock Holmes mystery; early-twentieth-century ones waited for comic books with their favorite action hero. Then film and television took over.

All storytelling follows Scheherazade's formula in *1001 Nights*: the storyteller survives not by completing a tale, making the point, and imparting a lesson, but by constantly setting up new, unfulfilled questions and anticipations. Beginning novelists are taught to withhold what readers want to hear most—what happens next. When an early chapter ends with a cliff-hanger, does the next one resolve the tension? Never. It's on to subplot B with other characters gaining our interest, creating an equally suspenseful situation. Then it's back to subplot A in the simplest structure, or on to C or D in more complex novels, leaving s hanging once again.

Entertainment has always functioned as a supernormal stimulus for social instincts, playing upon our urges to get to

Scheherazade and Shahryar.

know people and attend to compelling events. It is also an area where supernormal stimuli can potentially have supernormal payoffs. A great novelist constructs characters who act out a drama that will move us, teach us, and leave us better for the imaginary interaction than we'd be if we had spent the same amount of time interacting with those around us. But most entertainment is probably what Cascardi is calling "addictive"—less effortful but also less beneficial than real life.

The plots—and even *names*—of popular TV shows tell us which instincts they're tugging at. *Friends* brought into our livingroom a group of lively roommates, whose smiles, quips, and laughter caught us up in their camaraderie without our having to exercise any social effort. *Sex and the City* gave us more vicarious romantic adventures than we'd encounter in a lifetime. In both, svelte characters noshed constantly while never gaining weight; viewers joined in the eating from their own couches with quite a different outcome. *Cheers* took us to the ideal neighborhood gathering place where again we were guaranteed friendliness and humor no matter how lumpen our own behavior. If a contact high wasn't sufficient, we sipped along with a six-pack from our own fridge.

Innumerable hits from the 1950s—*Leave It to Beaver* and *Mayberry* through *Eight Is Enough* and *The Brady Bunch*—provided idealized parents with homey advice and perfect, agreeable youngsters. The list goes on: *Little House on the Prairie*, *All in the Family*, *Three's Company*, *Good Times*, and *Mad About You*. Each one's title promising fulfillment of a primal desire. Yet each one, after 30–60 minutes of the illusion of social contact, left you no richer in real friendship or family ties to support you in crisis and yet eager to tune in the following week to learn the next events in the lives of people who not only didn't care about you, *they didn't exist.*

Just as an orgasm with a flat image of another human being goes unquestioned in the modern world, so does any other kind of relationship with one. Whitedot.com does a good job of "making the ordinary seem strange" at their antitelevision site: "You are alone in the dark, staring at a plastic box. This is like a science fiction horror story." Of course you may not be alone—there may be friends or family in the next room . . . or at least potential friends just down the street. But the television pulls social instincts more strongly than these real people.

Attack of the Supernormal Stimuli

The other instinct television and films play to is that for adventure. For a plot to succeed without sex, romance, buddies, or family ties, there must be exotic travel, wild animals, car chases, or mountain climbing. It's no coincidence that "cliff-hanger" is a colloquism for a scene that captures our attention.

The viewer of adventure films vicariously explores exotic locales, creates empires, and struggles in confrontations with nature, man, and beast. The films are often set in romantic periods of the past and may feature real historic figures or literary heroes—Robin Hood, Tarzan, or Zorro, kings, battles, rebellion, or piracy. They involve expeditions for lost continents, jungle and desert epics, treasure hunts and medieval quests. They may have futuristic or interplanetary settings—all supernormally stimulating to our curiosity and instincts to explore. Adventure films target men more than women; males may have stronger instincts for physical adventure. However these themes appeal to women, too, and additionally they are often intermixed with subplots appealing to social instincts.

A related genre, the "action" film, puts more emphasis on continuous extreme activity—physical stunts, chase scenes, rescues, fights, and escapes—often in adversarial settings such as spy/counterspy or western cowboy/Indian. These are even more male-oriented. Films like *Guns of Navarone, The Wild Bunch*, and *Deliverance* follow all-male groups in grueling adventures. Two variations on the action film are war films and disaster films. War films stimulate many of the same instincts that leaders use to whip up real wars: vivid images of supernormal threats, portraying the enemy as a different pseudospecies. In the blockbuster war films that double as propaganda—the cold war films of the 1960s to 80s, the "war on terrorism" ones now—our side *always* wins. Unlike a real war in which we're perpetually asked to keep postponing our expectations of success, the film version of war rewards the tension with a clear victory.

Disaster films afford a unique window into our coded instinctive fears of the perils in our world. These films are never about a flu that kills 1 of 100 people (as the worst epidemic did), nor about a series of fatalities in small car crashes, or the planetary damage that will happen from global warming over a half century. As we discussed in the last chapter, to grab our attention a threat must be immediate and dramatic. And it's generally one that existed in the Stone Age—or at least can be easily extrapolated from something we feared then.

The list of disaster film titles reads as an inventory of our deeply instinctive fears: *Arachnaphobia, Ticks, Empire of the*

Ants, Mosquito, The Swarm, The Savage Bees, Anaconda, Snakes on a Plane, Jaws, Rabid, Infested, The Rats, Killer Rats, Plague, Outbreak, Virus, The Andromeda Strain, Avalanche, Tremors, Backdraft, Tornado, Twister, and *The Perfect Storm*. Some titles need images to connect them to a primal fear: *Cujo* is another rabies film, *Ben* and *Willard* are two more rat-attacks. Our Stone Age ancestors might never have feared a *Meteor*, but the film's poster makes clear it's simply the largest rock that's ever come hurtling toward you.

It may be embarrassing how many of these films we've not only heard of—or even seen—but also recall vividly. The mere mention of these concepts or the image of looming teeth or swarming vermin etch themselves into our memories. The effect may be most obvious for badly acted clunkers that we wouldn't watch if they didn't activate our alarm instincts. But consider great suspense films—Hitchcock's tales of murder, *Star Wars'* battles, or Indiana Jones's adventures. Our accolades of "classic" or "archetypal" basically mean that these films use acting, cinematography, and plot twists to push our instinctive buttons more strongly than any others.

Throughout history, man knew to pay attention when a rabid animal was on the loose or a fire raged out of control. Any mention of such a threat triggers an impulse to "go find out more about this." Those who thought, "Tornado . . . man-eating tiger . . . people dying of a mysterious disease? I don't care; I've got hunting and gathering to do," wouldn't have survived to leave many offspring. Our instincts won't let us ignore celluloid killer bees or rats any more than Tinbergen's sticklebacks could disregard the red-bellied dummy intruder.

Horror films ramp up the threat even more. What else are

King Kong, Godzilla, Frankenstein's monster, vampires, werewolves, aliens, radioactive mutants, zombies, or the Devil if not supernormal stimuli—exaggerated versions of the menacing human or wild animal? The pull of horror films is often explained as the thrill of an adrenaline rush; indeed, getting scared sometimes has this payoff because arousal of all types shares some common pathways. Roller coasters exploit this effect. But these films also end with the same clear victory as war films—Godzilla is defeated, the last radioactive ant killed. There is a ritual return to normality at the end.

However, it may not be safe to assume that all horror film audiences like the films in a simple sense. Some people who repeatedly view them experience negative aftereffects like nightmares, fear of the dark, and trouble sleeping. The supernormal fear stimulus doesn't necessarily make us happy—it makes us pay attention. As with the ordinary wild animal, the mention of a 20-foot-tall mutant evokes the "go find out more" response. This translates into ticket sales now, even if anxiety later. Many horror moviegoers describe something like an addiction: they hear of a new film premise, feel compelled to go to see it, become more anxious after seeing it, but somehow still feel compelled to see the next one.

We may have instincts telling us to pay attention when the Joker's taking over Gotham, or we've seen a poster about irradiated ants or mutant flesh-eating sheep (New Zealand's ultimate nightmare: *Black Sheep*). But this doesn't mean we can't override them. It means we need to use our brain to remind us we'd be wasting our time—as we sometimes but not often enough do. Even the people who experience vicarious satisfaction that their town hasn't been stomped by Godzilla come out of their hours of film or television viewing with their real-world challenges unresolved.

The very instincts that adventure and horror films cater to are the same ones that got our ancestors moving—hunting, exploring, practicing skills, . . . and fighting. At times of possible danger, they were what overcame the instinct we have telling us, "rest when you don't need to exercise"—a sound principle on the savannah when it was easy to get exhausted and hard to become too sedentary. If our ancestors saw something interesting, they got up to investigate. When the stimulus comes from the television, it invites us not to get up but to freeze and watch further.

George Zipf's *Human Behavior and the Principle of Least Effort* describes this in statistical rather than evolutionary terms. Across a wide variety of endeavors, people do what's easiest, not necessarily what's best. Zipf observed that most people, most of the time, are turned back by modest hurdles that they know could be overcome—with effort. "To be habitual, an action must be relatively effortless or carry a particularly large psychic reward," he observed. Zipf's "large reward" could be either instinctual or intellectually deduced and is definitely amenable to learning.[9]

Take Me Out to the Ball Game

Take me out to the ball game,
Take me out with the crowd;
Buy me some peanuts and Cracker Jack,
I don't care if I never get back.

—Jack Norworth, 1908

Postagriculture, when few people were hunting their own food or building shelter, most of the reward for movement came from sports and games—purely for fun or as jostling for dominance within a group or between groups. Tag and races developed running speed and endurance. Hide-and-seek provided practice for stalking prey and concealing oneself from enemies. Balls may have been early supernormal stimuli—thrown further and more predictably than other projectiles, rolling easier than the smoothest log—but games with these still developed strength, reaction time, and precision.

Those playing got some exercise—though team sports are generally not the optimal activity. But the biggest problem with most sports is that, as population densities grew, they quickly became spectator events—the impetus was channeled into vicarious experiences for the spectators. Until recently, many games had between a 1:1 and 1:3 ratio of players to spectators. The viewers often moved around the perimeter to get a better view of the action. And they certainly walked to the arena. The modern version of this subverts the "gaming instinct" yet further from exercise. Crowds drive to huge stadiums serving beer and hot dogs. Or they simply sit in their home recliner using the remote to switch between two games while gobbling chips and dip.

As with fictional film and TV, the vicarious experience of watching "real" sports gives viewers a false sense that they are engaging in activity and meeting challenges. A good part of the population, especially boys and men, spend hours of their week feeling like they're exercising while actually reclining motionless.

Aside from the inert spectators, the exercise for players

is often limited. "When I taught physical education and team sports," recalls Phil Lawler, director of the P.E.4Life Institute, "class lasted 42 minutes. We would take attendance; then teams were picked. The team put on tennis shoes and got on the field, gathered into a huddle, called a play, and went to the line of scrimmage. . . . When play began, 15 to 20 kids were standing around, doing nothing. We spent a lot of time arguing about who was offside."

Outside of school, the situation's no better. Little League dads and Soccer moms make a major commitment to their children's amateur sports careers. But much of their effort involves purchasing uniforms and driving to a game in a suburb on the far side of the city, where most of their time will be spent watching their kid sit on the sidelines. As one researcher on suburban sports noted, "During those 30 minutes that they actually play soccer each week, they may not spend even half the amount of energy that another kid spends walking to school."[10] And though they may think they're teaching lifelong skills, research shows that less than 5 percent of the population older than 24 uses team sports as a form of physical activity.[11]

Whatever happened to the impromptu neighborhood softball game? This institution provided quick exercise and plenty of practice forming teams and working out disagreements (the usual rationale for organized team sports). More so the "Go outside and play" when kids decided whether to climb trees, ride bikes, race each other, or play hide-and-seek—all affording practice of an array of social, athletic, and cognitive skills. And yet more in the spontaneous tumbling play and imitation of adults' activities that came naturally to children on the savannah. It's another instance of supernormal stimuli—uniforms, rituals, audience—replacing the real goal.

"Muscle Brain"

Growing up we probably all heard stereotypes about "dumb jocks" or "muscle brains." When an exercise equipment company mounted an ad campaign featuring muscular athletes with the caption: "Body by Nautilus," a popular rejoinder was "Brain by Mattel."

But these clichés have the relationship between fitness and cognitive abilities backward. Exercise *improves* alertness, thinking, and memory. The first group of mechanisms through which it does so are ones that benefit the entire body. Exercise stimulates the heart and cardiovascular system, improving circulation of blood within the brain. It raises the levels of oxygen and other nutrients and helps remove waste products generated by billions of neurons. Longer term, exercise reduces the levels of artery-clogging cholesterol, preserving the brain's blood supply just as it does elsewhere in the body.

Exercise also has more specific effects on the brain. Scientists once thought that mammalian brains stopped producing new cells early in life, but recent research indicates that we continue to manufacture new brain cells all our lives. This growth is increased in the face of novel learning situations or social interaction. However, the most potent stimulation of brain growth is . . . physical exercise.

In *Spark: The Revolutionary New Science of Exercise and the Brain*, psychiatrist John Ratey refers to exercise as

"Miracle-Gro."[12] Exercise increases levels of brain-derived neurotrophic factor or BDNF.[13] BDNF stimulates the production of new brain cells—neurons—and helps them survive.[14] Exercise especially generates neurons in the hippocampus, a part of the brain associated with memory, and these new neurons have been demonstrated to enhance learning.[15] Exercise increases dopamine and norepinephrine which facilitate attention, concentration, and help "lock in" memories as they form.

At all ages, physical activity improves learning and thinking. A 2002 California Department of Education study reported that the more physical-education classes students took, the higher their level of academic achievement.[16] Two large, long-term studies have found that physical activity in middle age and later years is associated with lower risks of cognitive impairment, Alzheimer's disease, and other forms of demen-

"I'm probably in the minority, but I would've loved to see Mantle on steroids."

tia.[17] Animal research demonstrates that exercise can increase neuronal survival after brain trauma,[18] and promote the growth of new blood vessels in the brain.[19]

In *Waistland*, I described the sea squirt, which starts life swimming with the aid of a brain and nervous system, but, after attaching itself to its permanent home, digests these now superfluous organs and pursues a vegetable-like existence. I used it only half in jest as a model of where humans might be heading. In fact some theories of consciousness suggest that brains evolved specifically for coordinating movement and that we cannot have thought without some covert motor activity.[20]

Nature's Medicine

Several recent studies report the "new finding" that exercise is an effective treatment for depression. Actually, this is old news. Thirty years ago, two studies established that running and other exercise programs helped clinically depressed patients recover.[21] Twenty years ago, running had been shown to be as effective as the class of antidepressants then most widely prescribed: the tricyclics.[22] A recent study reproduced this effect, comparing the use of stationary bikes and treadmills to Zoloft, one of the new serotonin-reuptake inhibitors—the same class as Prozac.[23] Six-month follow-up examinations revealed that subjects in the exercise group experienced lower relapse rates than those given the Zoloft. Reviews of the various studies of exercise for depression have concluded that exercise is effective both short and long term[24] and that aerobic activity, strength or flexibility training all prove effective in treating depression.[25]

The other effects of exercise—decreased risk of obesity, diabetes, heart disease, cancer, and Alzheimer's are well known. I only discuss them here in terms of why we're not more alarmed. We're bored to tears with hearing about the obesity epidemic—not scared into action—because it's not the kind of threat we've been discussing in past chapters as instinctively coded—immediate, easily visualized, caused by bad man or animal. When we hear about mutant giant ants, we head straight to the theater to find out more. The concept that we may die 15 years earlier than we need to of slow hardening of the arteries just doesn't grab our attention the same way.

Physiological researcher Frank Booth is the son of an advertising professional and therefore savvier than most academics about the importance of catchy names. He recently coined the phrase "Sedentary Death Syndrome." His Researchers against Inactivity-related Disorders (RID) publicizes the statistic that 250,000 deaths a year in the United States are caused by physical inactivity[26] and is lobbying the National Institute of Health to make exercise a higher priority. In fact, RID thinks exercise should be a component of every health study, pointing out that, because the human genome evolved within an environment of high physical activity, "the healthy 'control' group should actually be taken from a physically active population and not from a sedentary population with its predisposition to modern chronic diseases."[27]

Booth is not alone with this viewpoint. Sociologist Jeff Robbins catalogs how truly lazy we've become. During a rush hour one morning in New York City's Penn Station, he noticed a line of people waiting to take the escalator and few climbing the 18-step staircase. "One would think that they would choose to climb up the short flight knowing that getting the heart rate up with a little stair climbing would be good. They've been sitting for an hour or more on a train, or will, or are on their way to sit for seven, eight, or more hours at a desk typing, clicking, staring at a VDT."

Robbins bought a counter and returned for formal research. Excluding the disabled, anyone with children, animals, heavy bags, carts, skates, or bicycles, he counted 40,045 people who, for no apparent reason, took the escalator; only 5, 530 climbed the steps. One assumes *some* of theses people value exercise, but Robbins concluded, "The urge to minimize effort is so deeply ingrained, it's like air; we take it for granted; it fails to enter conscious thought."[28]

Supernormal Stimuli: The Next Generation

The 2008 film *WALL-E* takes place 700 years in the future. Mankind is living on a giant space cruise ship. In one sequence, the camera pans over portraits of the previous ship captains. We watch as successive generations gradually devolve into amorphous blobs. (CalorieLab's review questioned whether 700 years would really be needed for this to happen.)[29]

The "guests" spend their day in floating lounge chairs, eating and staring at chair-mounted video screens. One rolls off his chair and can't get back up. It takes only the arrival of a cute Disney robot alerting them to an enemy to mobilize these fictional humans into discovering the joy of exercise: "I didn't know there was a pool!" The enemy as motive for action is powerful, as we saw in the last chapter, but the desire for more exercise doesn't happen this easily. It takes more conscious reflection and willpower to reverse the slothful cycle and be rewarded by feeling better.

Other films play with disabusing us of our artificial social lives. *The Matrix* portrayed people floating in water-filled pods, plugged into computerized simulations so that they don't realize they're utterly alone. The protagonist is offered the chance to see the reality of his situation, then encouraged to leave his pod and explore the real world. It's ironic, of course, to sit motionless except for the occasional reach into the popcorn bucket while *WALL-E* prescribes diet and exercise or Morpheus warns us to abandon our illusory social world.

William James's prescription of making the ordinary seem strange should help us see all TV as odd and maladaptive. Even the news is hardly an efficient way to get information: we could read the same content much faster or listen to the radio while accomplishing physical tasks, if we even need that information. With a bit of questioning the ordinary, most TV news content seems very, very strange. Humans have instincts to communicate what is going on with other people and to pay attention to such messages. Almost everyone realizes that "news" of Britney Spears's drug and custody sagas is information that we don't really need despite instincts pushing us to learn more. Even the boy who's fallen down a well in another state is not something we're going to do anything about. Hearing of every girl who's lost in a forest or kidnapped by a sexual predator wildly raises our fears about these very unlikely risks. But even fatal falls, sexual abuse, and child murder are far likelier to happen in the home than in that great out of doors that we're more primed to fear. The far higher risks to our children rarely enter our minds: that they'll end up dying a decade before they need to of cardiovascular disease, diabetic complications, or cancer—in Booth's words, *of Sedentary Death Syndrome.*

Even "important" news might bear examination for how much of it really serves any logical purpose, like voting intelligently. Combing the planet for the largest, most dramatic disasters has some of the same effect that presenting fictional ones does and is somewhat of a supernormal stimulus itself. I wouldn't advocate totally ignoring world events. But if you're not one of the minority that sends money to earthquake victims, drives down to New Orleans or Galveston to rebuild houses, or takes effective antiwar action, what purpose does it serve to hear details of such events? While Britney and the child down the well are followed in a manner uncomfortably like soap operas, wars and natural disasters engage us more similarly to the disquieting but addictive horror films. Both steal time from smaller scale issues in our immediate environment on which we really could act.

Pundits had predicted that television viewing would drop as Internet use expanded. However, the reverse has happened. In 2008, the typical American watched 142 hours of television monthly, an increase of about 5 hours from the previous year. Internet use averaged more than 27 hours monthly, up an hour and a half.[30] On YouTube alone, 100 million viewers in the United States watch 6 billion videos each month.[31] People often can't bear to choose one screen over another: 31 percent of Internet use occurs in front of a television set.

If the Internet were replacing TV rather than adding screen

hours, it might have some slight advantages over television. While the Net has its share of mind-numbingly dumb content and a physical medium as passive as television, cognitively some of it is more interactive. Research on members of virtual reality sites reveals that activity correlates with more imagination, but this hasn't been controlled as carefully as most television data to tease out cause and effect.[32] Video gaming aids fine motor skills. One study found medical students who played more hours to have better initial surgical dexterity. Though, again, it's not clear which is cause and which effect.[33] Internet use, however, shares many of the same problems as television—correlating with attention deficit disorder and low frustration tolerance—not traits you'd want in that dexterous surgeon. And again, this is *in addition* to television viewing hours and separate from the 100 million Web- and video-equipped cell phones not yet counted in viewing hours.

The twentysomething son of one of my friends recently told me of finding himself without comprehensible television during a stay in a Japanese dorm. Most of his usual music, texting, and other media were not available either. Many evenings he was bored to tears. He resorted to the travel present given him by his father: a copy of Dostoevsky's *The Idiot*. At first, he found the paperback excruciatingly dull. He continued only because it was one notch better than doing nothing. Gradually, he became absorbed in the story of the epileptic prince and his complex romances, friendships, and rivalries. After reading its 600+ pages, he was awed at the subtle descriptions of passionately loving and hating the same person; of people being evil in some ways, admirable in others; and of the dilemma of loving two women at the same time. He never would have picked up the book—nor continued once started—if TV with its simpler, catchier, faster, laugh-track-accompanied stories had been available. But he did continue to read more classic literature upon his return home. We need to escape the super-normal pull of television to allow our instincts to be able to respond to subtler narratives.

Some of my generation would love to see the world get back to reading books, but there are two problems with this: (1) Books are about to disappear; and (2) They were already a somewhat artificial form. For those who disagree with the first point, as in "There will always be books; no computer screen is ever going to substitute for holding a leather bound volume in your hands," I'm sure earlier generations said, "No bound book is going to eliminate the pleasure of unrolling a papyrus scroll" or "The printing press can never produce a substitute for an illuminated hand-calligraphied volume."

It's a human drive to collect and pass along information and stories. The Internet is returning us a step closer to literacy than we were with television and music videos. Sites such as Project Gutenberg, loading public domain books online, and Wikipedia with its cooperative compilation of (mostly accurate) information show us what the Web can do. The user does, of course, have to type "Hamlet" into a search engine rather than "big boobs," "Beyonce download" or "Prada shoes." But a person who resolves to search for the wisdom of the ages can find it much more easily than a traveler on the savannah would have.

Screens will continue to get lighter and easier on the eyes. At first something like our present books will simply be read on these portable devices. But they're bound to evolve; newspapers already have transitioned to online versions. Electronic media allow interaction with the writer and other readers; this will—and should—be embraced. A video component easily integrates with words. Humans originally learned much of their information in narrative and visual forms. Computers already use synthesized voices to "talk." Verbal presentations will surely grow, perhaps along with miniature displays within eyeglasses or even directly wired into us as the most speculative futurists envision.

Our number-one effort should be to get back to more real social interactions. The relative merits of the new technologies are proportional to how compatible they are with this. Television interferes with all types of interactions. Couples with a set in their bedroom have half the sex and report less conversational intimacy than other couples. Family conversation correlates inversely with the amount of television watched. While television is simply a slower mode for adults to learn information, "educational television" for young children is a complete oxymoron. Toddlers learn by moving, vocalizing, interacting, and getting feedback. From any sort of TV, they learn to be passive.

Several "cold turkey" television withdrawal experiments have paid families to turn off their televisions for periods of a month or more. Others followed participants in the annual "turn off your television" week. These studies find that people experience adjustment problems only *very briefly*. One review of these studies summarized:

> The first three or four days for most persons were the worst. . . . In over half of all the house-holds, during

> these first few days of loss, the regular routines were disrupted, family members had difficulties in dealing with the newly available time, anxiety and aggressions were expressed. . . . People living alone tended to be bored and irritated.[34]

Family logs from the first few days of one study included the following entries: "The family walked around like a chicken without a head." "Screamed constantly. Children bothered me, and my nerves were on edge. Tried to interest them in games, but impossible. TV is part of them." And more ironically, "It was terrible. We did nothing—my husband and I talked."[35]

The longer-term effects, however, are remarkably beneficial. Within days, former television viewers adapt and rediscover their enjoyment of many other activities. Families play games, take walks, and talk more. Studies in schools enlisted parents in radically cutting back children's hours of viewing and found that both grades and physical fitness improved over the next year with no other intervention. We all need to do what these studies did—turn off our sets, get through the first few difficult days, and reset our attention spans to real interactions.

Computers present more ambiguous issues. Obviously they contain some great content. They are our new libraries, and they offer modes of arranging real social encounters. However, we need to offset their physical passivity and to avoid their more cognitively passive applications. Some computer gaming programs have begun to require significant exercise such as Nintendo's Wii Sports with programs asking the player to swing a tennis racket or bat as simulated landscapes and balls respond. Wii Fitness connects the computer to a balance board on which one performs aerobics, weight lifting, or yoga. Dance Dance Revolution comes with music and videos demanding a challenging sequence of steps on a footpad. Multiple companies make treadmills and stairclimbers to fit under workstations in place of a chair—even ones that require movement to activate the screen.

Likewise some social networking sites facilitate real social activities—help you to meet kindred souls you never would have found before the Internet. But with any technological activity, we need to consciously monitor whether it is serving our needs—social, informational, exercise—or whether it is introducing supernormal stimuli to hijack them. In the later case, we need to be quick to step away from the video game or Web site, and step out into the natural world.

8

Intellectual Pursuits as Supernormal Stimuli

Our brains did not evolve to do calculus or physics or to compose sonnets or fugues. And certainly not to work crossword puzzles or play chess—regardless of how many aficionados consider one of these to be the ultimate human achievement. Our brains evolved to solve survival problems on the savannah.

As our intellects grew more complex, humans were rewarded for seeking novel, challenging problems and attacking them with a passion that would have been futile for any other animal. This intellectual curiosity generates its own version of supernormal stimuli—problems more intriguing than any practical ones. The previous chapter discussed reading novels and watching film or television as lures for our social instincts. But the urge to create these art forms—or intellectual tomes like the present one—is also driven by instinct. Some of this too is social—the desire to express oneself and communicate—but the other component is intellectual—creating or solving a riddle to one's own satisfaction.

We've already discussed the evolutionary view of humans as a neotenous species of ape. And the domesticated fox experiments show that physical neoteny goes hand in hand with behavioral neoteny. Most animals are playful and curious only as juveniles and settle into set routines as adults. We humans show some change in this direction, but we retain more playful curiosity later in life than most species. This enables us to continue to learn and adapt at later ages.

Two hundred thousand years ago, as modern humans were emerging, hormones that triggered apes' maturation began to arrive later.[1] Growth hormone spurts that occur in the wombs of other mammals became postnatal, allowing the human head to grow larger. Sex hormones peaked several years after they had for chimps or early hominids, and this also allowed brain development to continue longer and behavior to remain more flexible. As mentioned in Chapter 4, this human characteristic is threatened by recent dietary changes and pollutants that are pushing our age of puberty back in the direction of our chimp relatives.

Novelty always provokes a combination of interest and fear, but changes in biochemistry can shift the balance. The corticosteroid surge associated with fear comes later for humans and never happens as strongly as for other primates—exactly the changes that make the domesticated foxes more curious and less fearful than their wild cousins.

More recently, genetic changes specific to brain development arose. Approximately 37,000 years ago a variant of the *microcephalin* gene appeared.[2] A rare, undesirable version of it makes brains smaller, hence the name. The new variant doesn't enlarge brains but seems to have subtler effects on their development. Evolutionary biologists are sure the new *microcephalin* gene conveys some advantage as it has been selected so strongly that 70 percent of the earth's people now carry it—as fast as genetic selection has ever been recorded. A clue to its function is that its spread tracks closely to the appearance of cave painting, the first musical instruments (reed flutes), and more sophisticated tool making.

A second brain-development gene, ASPM, mutated about 5,800 years ago and spread just as fast, now showing up in about 30 percent of humans.[3] Large cities and complex civili-

"You're certainly a lot less fun since the operation."

zations beginning with Mesopotamia tracked this gene, implying that it enables something not established by agriculture alone—perhaps it helps people adapt to agricultural densities. In fact, the only non-brain-linked mutation that has spread as rapidly is one that prolongs the ability of the digestive tract to process lactose—clearly useful for humans who began drinking cow's milk.

There hasn't been time in the last several thousand years for more genetic change, but society has begun to structure itself in ways that facilitate further neoteny.[4] Foremost among these is the prolonged duration of formal education. An ever-increasing proportion of the population graduates high school, attends colleges and universities, and goes on to graduate school. Universities reward youthful traits such as cognitive flexibility and the drive to acquire knowledge and skills. Higher education also delays key life experiences that tend to induce psychological maturity, such as marriage and parenthood.[5] Parenthood—especially pregnancy but even caring for young children by either parent—remodels the brain in detectable ways, so delaying this is yet another boost to neoteny.[6]

Brain Teasers

I'll get to the issue of how we use our giant brains in modern professions, but first I want to examine how we use them for fun. Mental play is the most clear, unbridled pursuit of supernormal stimuli for our neoteny, curiosity, and intellect. Humans have always poured enormous effort into solving puzzles and playing games. Perhaps the oldest, mazes—walled enclosures, hedges, stone patterns, or merely drawn on dirt or rock—have appeared across human history. Jigsaw puzzles have existed, at least in their wooden form, for about four centuries. Three-dimensional geometric puzzles have varied across time: Rubik's cube was an obsession for my generation, while Soma cubes were the equivalent a decade before that. These concrete puzzles hook the brain's desire to sort out configurations and patterns.

Other puzzles play to different intellectual challenges. Mathematical puzzles and magic squares call on our knowledge of numbers, visual illusions our spatial perception, and logic puzzles our abstract reasoning. Word puzzles include riddles, anagrams, ciphers, and codes. These challenge vocabulary, spelling, definition, and sound associations between words. They require us to play with metaphor and figure out tricky cues. We hardly need to do this for practical communication, but we're drawn in for hours at a time.

In 1913, English journalist Arthur Wynne published a "word-cross" puzzle in the *New York World.* The rectangular grid with squares to be filled in with cued words became wildly popular and its name was altered a few years later to "crossword."[7] Highbrow critics initially regarded the crossword puzzle as disdainfully as today's do the playing of video games. *The New York Times* was to become its most famous forum, but a 1924 editorial in that paper decried the "sinful waste in the utterly futile finding of words the letters of which will fit into a prearranged pattern, more or less complex. This is not a game at all, and it hardly can be called a sport . . . and success or failure in any given attempt

A chesslike Inuit game. Anonymous engraving.

is irrelevant to mental development.[8] A clergyman of the same period called the working of crosswords "the mark of a childish mentality"[9]—which is, of course, exactly what evolutionary psychologists would invoke to explain the puzzle's appeal to humans.

Games have a similarly long history with our species. The board game Go originated in China about 500 BC and called for a subtle balance of offensive and defensive strategies. Pachisi, now Anglicized in America as the familiar "Parcheesi," was created in India around 500 BC and utilized similar strategies combined with a larger dose of chance by letting dice determine some options.

Modern card games such as poker, bridge, or cribbage trace their exact rules back only a century or two, but the ancient Persian game of *As Nas* included key elements of betting, hand rankings, and bluffing. Dominoes, mahjong, and other tile games, which extend back centuries, call on quick pattern recognition. Charades and other cued guessing games have been played since hunter-gatherer times and involve social skills as well as intellectual inventiveness in depictions and interpretations.

Chess is perhaps the most popular game of all time. Its current form emerged in southern Europe during the second half of the fifteenth century, but it evolved from similar, much older games in India and Persia. Spurred on by chess's reputation as the most perfect illustration of the human intellect, in the 1960s computer scientists began to write programs that played chess. Many predicted that, before the sixties were over, a computer would be the world chess champion. By 1970, computer programs played at the strength level of an average high school chess player, and chess experts were emboldened to pontificate on how brute force searches could generate only routine play. Computers would never have the intuition or subtle intellectual skill to beat the best players, they predicted. By the early nineties, a computer defeated a chess master and then a grand master. In 1996, IBM's Deep Blue won a game against reigning world champion Gary Kasparov though Kasparov rebounded later in the tournament. The next year, Deep Blue won the rematch. Today, commercial programs running on home PCs achieve this level of play, and if you want to beat your computer, you had better take up kickboxing.

The Automaton Chess Player, also dubbed "The Turk," was a chess-playing machine first unveiled in 1770 for Empress Maria Theresa of Austria and exhibited continuously until its destruction by fire in 1854. Claiming to be a machine that could play chess formidably against a human opponent, The Turk was in fact an elaborate hoax—a human chess master hid inside, to operate the machine. With excellent operators, the automaton played and defeated most challengers including Napoleon Bonaparte and Benjamin Franklin. Although many had suspected the hidden human, the hoax became widely known only in the 1820s after Londoner Robert Willis published his *An Attempt to Analyse the Automaton Chess Player* (London, 1821). Source: Karl Gottlieb von Windisch, *Inanimate Reason* (London: S. Bladon, 1784).

Computers are used for much game play. Tetris, the most popular computer game of all time, is a cross between two-dimensional building blocks and a jigsaw puzzle swirling around in electronic space. Lexulous, currently the most popular game on Facebook, is a slight variant on Scrabble. (In fact, it was called Scrabulous until Scrabble's copyright holder,

Hasbro, sued.) Often the computers are simply an interface enabling players to compete with each other on the Internet—Lexulous and most card-game sites are like this. Some online versions of Scrabble and other games allow you to play the machine itself. Just as with chess, these programs could always beat the hell out of a human player, but they usually feature handicapping levels to dumb themselves down.

It's mainly when games are new that critics grumble about how much time these pursuits waste. We hardly remark on adults spending numerous evenings at the old standbys chess or Sudoku or traveling halfway round the world for bridge or Scrabble tournaments. Soon marathon Lexulous games on Facebook will be as respectable. But that doesn't mean the games cease diverting energy from other tasks.

Research occasionally reports some benefit from games. I've already mentioned the study finding that video games enhanced surgical speed. Another reported that people who worked crosswords averaged a later onset of Alzheimer's disease, but the direction of cause and effect in such studies is not entirely clear. Even if the games have a benefit compared to another use of time, the comparison group probably isn't using their extra hours practicing work skills or developing better social relationships. More likely, the studies reflect the damning-with-faint-praise effect: that games are better for you than . . . television.

Nonetheless, games and puzzles are the clearest examples of supernormal stimuli for our intellectual curiosity. Seeing a puzzle and impulsively needing to solve it is like seeing a cute toy and wanting to nurture it, like seeing a paper centerfold and wanting to have sex with it, or hearing of an exaggerated threat and wanting to fight it. Keep this effect in mind as we explore what proportion of our professional pursuits are shaped by the same reflexive responses.

Supernormal Careers

Many fields seem powered by intellectual curiosity rather than immediate, practical application. Universities are filled with people studying abstract questions of philosophy and aesthetics. Even practical science often begins with someone wondering what those things shining in the sky are, how the body works, or what constitutes the smallest building block of matter. Many discoveries described in this book arose by people trying to deduce our ancestors' habits by examining bones and artifacts or trying to visualize what the earth was doing earlier yet by searching through layers of sediment. Darwin was motivated by something similar when he came up with his theory of evolution. Tinbergen was endlessly curious about how animal societies were organized.

In the idealistic 1960s, people grew concerned about the dangers of nuclear reactors. Solar, geothermal, and wind power were touted as the safer options of the future. I remember my physicist father telling me something that proved remarkably true. He said nuclear energy would continue to develop while natural energy sources lagged behind for the same reason they were already discrepant: nuclear physics was *interesting*. Fascinating challenges about the nature of subatomic particles and how they could be manipulated remained open. The collection of solar, wind, or geothermal energy was basically "fancy plumbing." The delay wasn't that their unsolved problems were too difficult. They were *too easy*. Or at least dull: practical details of materials, cost, manufacturing. Work on a solar panel or wind turbine wasn't going to illuminate any secrets of the universe.

Indeed, over the intervening decades, scientists and governments have continued to build and elaborate nuclear reactors while work on natural energy sources has been slow. The $6+ billion spent on the Large Hadron Collider alone is more than the sum total of all research developing the fancy plumbing of natural energy. The collider searches for bosons, fermions, and other new particles predicted by the theory of supersymmetry. These probably won't turn clean energy generators—it's unclear what they will do since it's still unclear if they exist. But they're really, *really* interesting.

More typical professional lives are also determined by the pull of neotenous curiosity. In the third world, of course, many people do whatever work allows them to survive. In our society, some choose a job based on earning the largest paycheck. Others follow social instincts—altruistic ones about service to others or hedonistic desires to work in the liveliest social setting. But many people seek what interests them intellectually, as high schools and colleges explicitly encourage them to do. This is obvious with intellectual careers: Does manipulating numbers intrigue you? Finding the best wording to express something? Writing computer code? But it's also true with the trades: Like to tinker with the motors of cars? Grow beautiful plants? Concoct new recipes? When starvation isn't an issue, people gravitate toward their favorite professional puzzles.

Art is the field in which curiosity and play have the most obvious role. Artists generate supernormal stimuli for aesthetic ideals. Musical instruments and song refine and amplify tones that signal health and friendliness. Painting and sculpture capture the images we find most beautiful—naked human bodies, expressive faces, lush fruit, flowers, and landscapes. Impressionistic and abstract art play with colors and forms extracted from these. The stereotype of the artist—with at least some grain of truth—is as the perpetual adolescent, eluding the usual responsibilities of adulthood. Artists are likelier to receive gratification from play and creativity, and perhaps fame or admiration, than from traditional accumulation of wealth or power.

At the other end of the continuum, those in the world of business make it a point to go after money explicitly along with all of Chapter 5's drives for property and possessions. When you look beneath the surface of this milieu, however, you find people fascinated by intellectual puzzles—the creativity of investment strategies, mathematical formulas for tracking economic trends, intuitions that keep them one step ahead of their rivals. The popular adage, "Money is just how you keep score" isn't a random expression. It reveals much about the motivation as game or play—how close Wall Street is to chess or Scrabble. The recent recession has reminded us that our stock certificates, commodities futures, and even money don't map precisely onto concrete possessions.

Despite our instincts to claim yet more objects, land, and possessions, the wealthy and powerful no longer have more offspring. Eight hundred years ago, Genghis Khan conquered so much territory that he ruled the largest contiguous empire in the history of the world. He had more wives and concubines than any man of that era, and historians also think he raped many women taken prisoner by his armies. Modern DNA researchers find that nearly 8 percent of the men in the area of the former Mongol empire carry identical copies of a Y chromosome believed to be Khan's. That translates into 0.5 percent of the men in the world, or roughly 16 million descendants.

Modern rulers do not leave more offspring than anyone else, even leaders like JFK who are rumored to have had scores of sexual partners. The uncoupling of wealth and power from genetic proliferation is due to one factor: the advent of simple, effective birth control. Nature endowed us with an overwhelming desire to have sex, but with only a weaker, quirky, and intermittent drive specifically to have children—and rarely *lots* of children. Now that sex doesn't lead automatically to offspring, other instincts—to avoid hurting one's primary partner, to avoid extensive responsibilities—come into play and conflict with maximum procreation.

Survival and fitness may not have just separated, their relationship may even have reversed in recent history. The controversial 1994 *The Bell Curve*[10] included dubious assertions about ethnicity and social policies, but outlined research indicating one indisputable fact. People doing less well by most criteria—IQ, years of education, money earned, a stable family unit, and the like—now produce the most offspring. In the developed world at least, the vast majority of children who are born will survive. If their parents aren't able to provide for them, people with more resources will contribute or outright adopt them. The offspring of the less successful survive to reproductive age, and pass along their genes at a faster pace than anyone else. *The Bell Curve* didn't mention the cuckoo, but it's an apt metaphor for the book's main point.

Instincts for reproduction also combine with the intellectual curiosity discussed in this chapter to direct our resources in more exotic directions.

Intelligent Design

"Many people have said to me that [building robots] is something men do because men can't have babies themselves," says MIT roboticist Rod Brooks. "But it turns out, in my lab, there are more women than men. I think there is some deeper-seated thing which crosses sex boundaries about understanding life by building something that is life-like."[11] Early robotics was rife with optimism that artificially intelligent machines would come to surpass existing life-forms in intelligence and durability. Some calmly predicted that humans would go down in history mainly as providing the evolutionary bridge from carbon-based life to silicon-based species.

Such forecasts have scaled back, not because robotics looks less promising, but because biology has caught up in its ability to engineer carbon-based life. Artificial life-forms now seem likelier to be humans equipped with silicon add-ons enhancing strength or information processing. Purely carbon-based life with amazing new abilities seems equally plausible given the burgeoning field of genetic engineering.

Cloning has received the most publicity because its technology is already here. Ever since 1996 when Dolly the Sheep was cloned, the possibility of doing the same with humans has been hotly debated. Dolly's creators, Ian Wilmut and Keith Campbell, were quick to say they wouldn't consider cloning a human and found it "repugnant in general."[12] For a while, biologists thought that human cloning might be impossible because primate DNA divides so differently. The simple nuclear transfer technique used in Dolly did not work for monkeys. But in 2001, Advanced Cell Technologies cloned a four-cell human embryo. In 2008, Stenmagen cloned a 100-cell blastocyte, or mature embryo. Both companies were quick to say they didn't know if their embryos were viable to grow to birth and that they would never try implanting one in a womb. *The New York Times*, however, quoted a leading fertility specialist saying, "We've seen reproductive blastocysts that look like this or worse and they implant."[13] All major scientists working with cloning continue to deny this would ever be considered for humans.

But behind closed doors? Wealthy people who've lost children contact cloning facilities and ask to have their child cloned. They're turned away. But it would take $25,000 of equipment. Many fertility clinics are equipped with someone who might be able to do this. When might it actually happen?

Three years ago, the Cold Spring Harbor Laboratory on Long Island hosted a summit of cloning experts. Almost everyone who'd participated in cloning any species was there, including Dolly's creators. No one other than invitees was allowed in, and there were no media, no tape recorders. However, Michael Bishop, president of Infigen, later told a *Wired* reporter of one especially interesting conversation. "One evening after dinner, some of us were talking [about human cloning], and there was not one of us who believed it had not already happened. . . . It is being done," he stated. "I have no doubt. It would be stupid and naive to think it's not."[14]

If we begin to create embryos through technology, however, we needn't be bound by either a mother/father set of chromosomes nor the exact DNA of one individual. A process known as "homologous recombination" could enable scientists to remove undesirable traits and replace them with desired ones, one gene at a time. This has already been done in an experiment with mice which won the 2007 Nobel Prize in Medicine.[15] A speedier route would be to introduce a group of traits together by inserting an entire new chromosome. "Based on what we know, the artificial chromosome is going to be the best way to modify the genome," says Lee Silver, a professor of molecular biology and public policy at Princeton University. "Nature doesn't care about individual children. Instead of rolling the dice, why don't we take the dice and put them down in the way that parents think is best for their children."[16] Silver anticipates the development of specialized artificial chromosomes—a "good health" one, for example—that could routinely be inserted into human embryos to give them lifelong protection against cancer, strokes, and heart disease.

"I was wondering when you'd notice there's lots more steps."

Some biologists predict that gene targeting and artificial chromosomes could leapfrog over evolution and let us take control of our genome, maybe even turn ourselves into a whole new species. "There is no scientific basis for thinking that we couldn't," Silver says. "There's nothing really special about the human genome. There's nothing that says this is the end."[17]

Given what we've seen with other scientific ideas like nuclear energy, if the technology is interesting, someone will pursue it. This would give us intentional design. We might want to think about making it intelligent design.

The traits futurists usually discuss engineering are beauty, height, athleticism, creating superwarriors or people who can eat all the calories they want and not get fat or sick. These are all goals our instincts push us toward that would essentially enable further supernormal stimuli. The futurist movement seems still in the grip of present and even past goals—how to pursue defunct hunter-gatherer agendas even more tenaciously. What's rarely mentioned is altering instincts themselves. How about designing people who won't fight or don't crave excess fat or sugar? We may well not want to engineer human beings. But if we do, speeding up what natural selection would do through many millennia of suffering would seem to be the way to go: retooling desires to fit the new environment, getting rid of drives that no longer serve survival.

Changing instincts may sound incredibly complex, but no more so than any other genetic engineering. Of course, none of this is going to be possible tomorrow. And we don't need these high-tech interventions to get at the essence of the problem. How we would apply them has a lot in common with reining in instincts with the means currently at our disposal.

Conclusion: Get Off the Plaster Egg

In our generation, we need to begin to engineer our environment back to something more like what we were designed for and also to notice and resist whatever supernormal stimuli inevitably remain around us. Collectively we can decide to make our environment more walkable, to tax or even ban junk food, to reduce televisions blaring around us. We can explicitly teach that pseudospecies are illusory: we can broadcast the opinions, similarity, humanity of people across the world, train people to be automatically wary of leaders asking us to fight another group of humans pretty much like ourselves.

Individually, we must first identify supernormal stimuli. We don't have to just "listen to our instincts," we can exercise willpower—almost a dirty word these days, but a trainable skill shown to help habitual problems.[1] That's what our giant brains were designed for—overriding reflexive instincts when they start to lead us astray.

In a world increasingly designed to stimulate hunger, sexual arousal, and acquisitiveness, chasing the supernormal is a losing game. It's not antihedonistic to rein in, or redirect, instincts. Our pleasure system is robust and *very* flexible. Scientific studies show that people experience similar levels of happiness long term regardless of external events. People in the poorest nations are a couple percentage points less happy than those in the most affluent.[2] People who drink, don't drink, watch TV, don't watch TV, eat natural vegetables, eat junk food—all experience similar levels of life satisfaction. In fact, the only thing that makes a difference is chronic pain or consistent health crises—things our modern pursuit of supernormal stimuli tend to produce.[3]

The pleasure mechanism can be shaped as to what it responds to. It doesn't have to be the other way around. We understand this more readily when we're thinking about the evanescent highs of a drug-addicted life versus pleasure of normal life interactions. But the same is true for diet or social activities or what you find cute. People get pleasure from what they have gotten used to getting pleasure from. Reward circuits in our brain will respond to the sugar in a handful of tart berries or from the whole pie, to the earned rest after exercise or the whole day on the couch, to real friends dropping by or the simulated laugh track of a sitcom—all depending on what habits we get accustomed to.

The key to most of our modern crises lies in "making the ordinary seem strange." We are the one animal that can notice, "Hey, I'm sitting on a polka-dotted plaster egg" and climb off.

Acknowledgments

Fellow writers Ellie Tonkin and Andrew Szanton valiantly edited the turgid academese of my early drafts. My friends and colleagues Morty Schatzman and David Spiegel lent their encyclopedic knowledge of psychology, biology, and medicine to help me refine my thesis. Don Symons, with his intimate knowledge of human sexuality, was the perfect critic for the sex chapter. I'm grateful to Ruth Lingford and Olga Michnikov for everything from practical suggestions about images to hugs and cups of tea as they patiently listened to more than they ever wanted to know about supernormal stimuli.

Members of my two writing groups contributed invaluable suggestions and encouragement as well as sharing the inspiration of their own work. Thanks to Judah Leblang, Gary Simoneau, Harlow Robinson, Janet Spurr, James Tobin, Alyce Getler, Beth Rider, Ayse Atasoyla, Patti Heyman, and Ruth Cope. I'm indebted to my editor at Norton, Angela von der Lippe, for her work with me on three books at two publishing houses over the last fifteen years. Her assistant, Erica Stern, shepherded *Supernormal Stimuli* into print with unflagging attention to detail.

My thesis builds on the ideas and work of so many people. Some are quoted prominently throughout, including Niko Tinbergen, Konrad Lorenz, and William James. Others are mentioned only in the myriad reference notes at the end. The book—and indeed society—is indebted to these hundred of researchers in ethology, sociobiology, and evolutionary psychology who are helping explain the enigmas of human behavior.

Notes

Chapter 1: What Are Supernormal Stimuli?

1 *Milton Quarterly* 39, no. 4 (2005), pp. 271–73.

2 Seth Borenstein, "Marked-Up Birds Become Chick Magnets," AP Science Newsfeed, 2008-06-04 08:16:30.

Chapter 2: Making the Ordinary Seem Strange

1 Niko Tinbergen, "Autobiography" at the Nobel Prize Web site: http://www.nobel.se/medicine/laureates/1973/tinbergen-autobio.html (accessed December 15, 2008).

2 Hans Krunk, *Niko's Nature* (Oxford, England: Oxford University Press, 2003).

3 Krunk, *Niko's Nature*, pp. 29–30.

4 N. Tinbergen, Voorjaar [Spring] *De Meidoorn* 1 (1929), pp. 56–58, trans. in Krunk, p. 31.

5 Krunk, *Niko's Nature*, p. 56.

6 Krunk, *Niko's Nature*, p. 58.

7 Auke R. Leen, "The Tinbergen Brothers" at the Nobel Prize Web site: http://nobelprize.org/nobel_prizes/economics/articles/leen/index.html (accessed December 15, 2008).

8 Krunk, *Niko's Nature*, p. 90.

9 Desmond Morris, *Animal Days* (London: Jonathan Cape, 1979).

10 Konrad Lorenz, "My Family and Other Animals," in *Studying Animal Behavior*, ed. D. Dewsbury (Chicago: Chicago University Press, 1985), pp. 259–87.

11 Krunk, *Niko's Nature*, p. 117.

12 Alec Nisbett, *Konrad Lorenz* (New York: Harcourt Brace Jovanovich, 1976), p. 97.

13 Krunk, *Niko's Nature*, p. 127.

14 Konrad Lorenz, *King Solomon's Ring*, trans. Marjorie Kerr Wilson (London: Methuen, 1952). First published in German, 1949.

15 Krunk, *Niko's Nature*, p. 130.

16 Krunk, *Niko's Nature*, p. 202.

17 Krunk, *Niko's Nature*, p. 131.

18 Desmond Morris, *The Naked Ape* (New York: McGraw-Hill, 1967).

19 Richard Dawkins, *The Selfish Gene* (New York: Oxford University Press, USA, 1976).

20 Krunk, *Niko's Nature*, p. 263.

21 Krunk, *Niko's Nature*, p. 260.

22 P. Marler and D. R. Griffin, "The 1973 Nobel Prize for Physiology or Medicine," *Science*, November 2, 1973, pp. 464–67.

23 Richard Dawkins, *The Selfish Gene* (new ed.) (New York: Oxford University Press, 1989), p. 1.

24 Charles Darwin, *The Expression of the Emotions in Man and Animals* (New York: D. Appleton, 1896).

25 Konrad Lorenz, *Civilized Man's Eight Deadly Sins* (New York: Harcourt Brace Jovanovich, 1974).

26 E. O. Wilson, *Sociobiology: The New Synthesis* (Cambridge, MA: Harvard University Press, 1975).

Chapter 3: Sex for Dummies

1 Kurt Vonnegut, *God Bless You, Mr. Rosewater* (New York: Delacorte Press, 1965), p. 109.

2 Warren Ellis, "Second Life Sketches: Please Stop Doing That to The Cat," Reuters, February 23, 2007, and reader response posted at Reuters news blog, February 28, 2007.

3 Shere Hite, *The Hite Report on Male Sexuality* (New York: Alfred A. Knopf, 1981). See the appendix for a statistical breakdown of findings (p. 1123): Do you look at pornography? (results for 7239 men): Regularly, 36 percent; Sometimes, 21 percent; Infrequently, 26 percent; Used to, 6 percent; No, 11 percent.

4 Alison King, "Mystery and Imagination: The Case of Pornography Effects Studies," in *Bad Girls and Dirty Pictures*, ed. Alison Assiter and Avedon Carol (London: Pluto Press, 1993).

5 Milton Diamond, "Pornography, Rape and Sex Crimes in Japan," *International Journal of Law and Psychiatry* 22, no. 1 (1999), pp. 1–22.

6 Cato Institute, "Do Movies and Music Cause Violence? Sex, Cyberspace, and the First Amendment," Cato Institute Policy Report, January/February 1995.

7 Diamond, "Pornography, Rape and Sex Crimes."

8 Joseph W. Slade, "Violence in the Hard-Core Pornographic Film: A Historical Survey," *Journal of Communication* 34, no. 3 (1984), pp. 148–63.

9 Gershorn Legman, *Love and Death: A Study in Censorship* (New York: Breaking Point, 1949).

10 Avedon Carol, "Snuff: Believing the Worst," in *Bad Girls and Dirty Pictures*, ed. Alison Assiter and Avedon Carol (London: Pluto Press, 1993); and Eithne Johnson and Eric Shaefer, "Soft Core/Hard Core: Snuff as a Crisis in Meaning," *Journal of Film and Video* 45, no. 2-3 (1993), pp. 40–59.

11 Donald Symons, *The Evolution of Human Sexuality* (New York: Oxford University Press, USA, 1981). Pages 166 to 196 contain a discussion of many of the earlier of these studies. Despite the intervening two decades, this classic book remains the best at articulating the evolutionary psychology perspective on sexual issues.

12 J. H. Langlois and L. A. Roggmann, "Attractive Faces Are Only Average," *Psychological Science* 1 (1990), pp. 115–21.

13 S. W. Gangestad and D. M. Buss, "Pathogen Prevalence and Human Mate Preferences," *Ethology & Sociobiology* 14 (1993), pp. 89–96; and S. W. Gangestad, and R. Thornhill, "Menstrual Cycle Variation in Women's Preference for the Scent of Symmetrical Men," *Proceedings of the Royal Society of London* 265 (1998), pp. 927–33.

14 V. S. Johnston and M. Franklin, "Is Beauty in the Eye of the Beholder?" *Ethology and Sociobiology* 14, no. 3 (1993), pp. 183–99.

15 For more about ideal weight, see Deirdre Barrett, *Waistland: The R/Evolutionary Science Behind Our Weight and Fitness Crisis* (New York: W. W. Norton, 2008), especially pages 111–59.

16 Jayne Krentz (ed.), *Dangerous Men and Adventurous Women* (Phila-

delphia: University of Pennsylvania Press, 1992), for lots of these characterizations.

17 Gorry statistics quoted in Salmon and Symons, *Warrior Lovers* (London: Weidenfeld & Nicolson, 2001).

18 Salmon and Symons, *Warrior Lovers*, p. 69.

19 Over 64 million people claimed to have read at least one romance novel in 2004, according to a Romance Writers of America study, a 26 percent increase over their 2001 study. Romance readers were split evenly between people who were married and those who were single. "Romance Writers of America's 2005 Market Research Study on Romance Readers," https://www.rwanational.org/eweb/docs/05MarketResearch.pdf (retrieved April 16, 2007).

20 Janice Radway, *Reading the Romance: Women, Patriarchy, and Popular Literature* (Chapel Hill: University of North Carolina Press, 1984).

21 Liz Kelly, "The First Crush Is the Deepest," Celebritology at http://blog.washingtonpost.com/celebritology/2008/01/friday_list_the_first_crush_is.html. All responses are from this *Post* blog, but the tallies and percentages are my calculations from it.

22 H. Fisher, *Anatomy of Love: The Mysteries of Mating, Marriage, and Why We Stray* (New York: Simon & Schuster, 1992), p. 106; and Jared Diamond, "The Worst Mistake in the History of the Human Race," *Discover Magazine*, May 1987, pp. 64–66.

23 E. Westermarck, *The History of Human Marriage* (London: Macmillan, 1891).

24 Fisher, *Anatomy of Love.*

25 Geoffrey F. Miller, "Evolution of the Human Brain through Runaway Sexual Selection: The Mind as a Protean Courtship Device," unpublished thesis, pp. 237–38 for this quote. For more on this perspective generally, see chapters by Geoffrey Miller and by Mary Hotvedt in *Mating Intelligence: Sex, Relationships, and the Mind's Reproductive System*, ed. Glenn Geher and Geoffrey Miller (Mahwah, NJ: Lawrence Erlbaum Associates, 2007).

26 Symons, *Evolution of Human Sexuality*, p. 108.

27 A. H. Schultz, *The Life of Primates* (London: Weidenfeld and Nicolson, 1969).

28 Alfred Kinsey et al., *Sexual Behavior in the Human Male* (Philadelphia:

W. B. Saunders, 1948). These researchers actually found 10 percent in the 1950s in a population possibly skewed by willingness to talk with researchers.

29 Witness tigons, ligers, leopons, zebra-horse hybrids, wholphins, and a host of animals either never or rarely occurring in nature. See J. A. Coyne, "The Genetic Basis of Haldane's Rule," *Nature* 314, no. 6013 (1985), pp. 736–38 for a summary of how hybridization occurs.

30 For discussions of this in a number of countries, see Peter Tatchell, "Is Fourteen Too Young for Sex?" *Gay Times*, June 1996, pp. 36–38; Peter Tatchell, "Why the Age of Consent in Britain Should be Lowered to Fourteen," *Legal Notes* 38 (London: Libertarian Alliance, 2002); I. Lucas, *Outrage! An Oral History* (London: Continuum, 1998), pp. 214–15, cited in Matthew Waites, *The Age of Consent—Young People, Sexuality and Citizenship* (New York/London: Palgrave Macmillan, 2005), p. 220. Francis Bennion, *Sexual Ethics and Criminal Law: A Critique of the Sexual Offences Bill 2003* (Oxford: Lester Publishing, 2003), p. 13, cited in Waites, pp. 220 and 222). Miranda Sawyer, "Sex Is not Just for Grown-ups," *The Observer*, Review section, November 2, 2003, pp.1–2.

31 BBC Channel 4, "Sex Before 16: Why the Law Is Failing" November 16, 2003. Thirty-four percent thought the age of consent should be reduced to 14, 35 percent thought it should stay at 16, 13 percent thought it should be raised to 18, and 18 percent thought it should be abolished.

32 F. J. P. Ebling, "The Neuroendocrine Timing of Puberty," *Reproduction* 129, no. 6 (June 1, 2005), pp. 675–83; A.-S. Parent et al., "The Timing of Normal Puberty and the Age Limits of Sexual Precocity: Variations around the World, Secular Trends, and Changes after Migration," *Endocrinology Review* 24, no. 5 (October 1, 2003), pp. 668–93; K. K. Davison, E. J. Susman, and L. L. Birch, "Percent Body Fat at Age 5 Predicts Earlier Pubertal Development Among Girls at Age 9," *Pediatrics* 111, no. 4 (April 1, 2003), pp. 815–21; P. Kaplowitz, "Clinical Characteristics of 104 Children Referred for Evaluation of Precocious Puberty," *Journal of Clinical Endocrinology and Metabolism* 89, no. 8 (August 1, 2004), pp. 3644–50; S.-Y. Ku et al., "Age at Menarche and its Influencing Factors in North Korean Female Refugees," *Human Reproduction* 21, no. 3 (March 1, 2006), pp. 833–36; K. Casazza, M. I. Goran,

and B. A. Gower, "Associations among Insulin, Estrogen, and Fat Mass Gain over the Pubertal Transition in African-American and European-American Girls," *Journal of Clinical Endocrinology and Metabolism* 93, no. 7 (July 1, 2008), pp. 2610–15; J. M. Lee et al., "Weight Status in Young Girls and the Onset of Puberty," *Pediatrics* 119, no. 3 (March 1, 2007), pp. e624–30; Paul B. Kaplowitz et al., "Earlier Onset of Puberty in Girls: Relation to Increased Body Mass Index and Race," *Pediatrics* 108, no. 2 (August 2001), pp. 347–53.

33 Y. Wang, "Is Obesity Associated with Early Sexual Maturation? A Comparison of the Association in American Boys Versus Girls," *Pediatrics* 110, no. 5 (November 1, 2002), pp. 903–10; J. M. Kindblom et al., "Pubertal Timing Is an Independent Predictor of Central Adiposity in Young Adult Males: The Gothenburg Osteoporosis and Obesity Determinants Study," *Diabetes* 55, no. 11 (November 1, 2006), pp. 3047–52.

34 Areeg H. El-Gharbawy et al., "Serum Brain-Derived Neurotrophic Factor Concentrations in Lean and Overweight Children and Adolescents," *Journal of Clinical Endocrinology and Metabolism* 91, no. 9 (2006), pp. 3548–52.

Chapter 4: Too Cute

1 K. A. Hildebrandt and H. E. Fitzgerald, "Facial Feature Determinants of Perceived Infant Attractiveness," *Infant Behaviour and Development* 2 (1979), pp. 329–39; K. A. Hildebrandt and H. E. Fitzgerald, "The Infant's Physical Attractiveness: Its Effect on Bonding and Attachment," *Infant Mental Health Journal* 3 (1983), pp. 3–12; T. R. Alley, "Infantile Head Shape as an Elicitor of Adult Protection," *Merrill-Palmer Quarterly* 29 (1983), pp. 411–27; V. McCabe, "Facial Proportions, Perceived Age, and Caregiving," in *Social and Applied Aspects of Perceiving Faces*, ed. T. R. Alley (Mahwah, NJ: Lawrence Erlbaum Associates, 1988), pp. 89–96.

2 *Eye of the Leopard*, National Geographic HD Channel, December 17, 2006.

3 Adriana Silvia Benzaquen, *Encounters with Wild Children: Childhood, Knowledge, and Otherness* (Toronto, Canada: York University Press, 1999).

4 BBC Channel 1, "Living Proof," October 13, 1999.

5 John McCrone, "Wolf Children and the Bifold Mind," *The Myth of Irrationality: The Science of the Mind from Plato to Star Trek* (New York: Carroll & Graf, 1994); Joseph Amrito, Lal Singh, and Robert M. Zingg, *Wolf-Children and Feral Man* (North Haven, CT: Shoe String, 1966).

6 Several relevant studies are summarized in R. Mazella and A. Feingold, "The Effects of Physical Attractiveness, Race, Socioeconomic Status, and Gender of Defendants and Victims on Judgements of Mock Jurors: A Meta-Analysis," *Journal of Social Psychology* 24 (1994), p.1315.

7 History World, http://www.historyworld.net/wrldhis/PlainTextHistories.asp?groupid=1813&HistoryID=ab57 (retrieved October 18, 2005).

8 Lyudmila N. Trut, "Early Canid Domestication: The Farm Fox Experiment," *American Scientist* 87, no. 2 (March-April 1999).

9 J. Lindberg et al., "Selection for Tameness Has Changed Brain Gene Expression in Silver Foxes," *Current Biology* 15, no. 22 (2005), pp. R915–16.

10 Stephen Jay Gould, "A Biological Homage to Mickey Mouse," in *The Panda's Thumb: More Reflections in Natural History* (Harmondsworth, Middlesex: Penguin Books, 1980), p. 89.

11 Lorenz, *The Foundations of Ethology*, p. 164.

12 R. A. Hinde and L. A. Barden, "The Evolution of the Teddy Bear," *Animal Behaviour* 33 (1985), pp. 1371–73.

13 P. H. Morris, V. Reddy, and R. C. Bunting, "The Survival of the Cutest: Who's Responsible for the Evolution of the Teddy Bear? *Animal Behaviour* 50, no. 6 (1995), pp. 1697–1700.

14 Mary Roach, "Cute Inc," *Wired*, no. 7 (December 1999).

15 Merry White, *The Material Child: Coming of Age in Japan and America* (Berkeley: University of California Press, 1994).

Chapter 5: Foraging in Food Courts

1 Charlie LeDuff, "Step Right Up, Ladies and Gents, to See the End of an Oddity," at http://travel.nytimes.com/2006/11/13/us/13album.html (retrieved November 13, 2006).

2 P. Patton, "America's Ever-Bigger Bottoms Bedeviling Seating Planners," *Miami Herald*, September 23, 1999.

3 If airlines don't want to use the new estimates they can either weigh

passengers (none have opted for this) or ask them their weight *and add 10 pounds to this*—the average underestimate in FAA studies. Mathew Wald, "F.A.A. Reviews Rules on Passenger Weight After Crash," *The New York Times*, January 28, 2003.

4 Daniel Gross, "Gross National Product: Obesity Spurs U.S. Economic Growth," *Slate* July 21, 2005.

5 BBC News, "Crematoria Struggle with Obese" (p.d. April 18, 2007) at http://news.bbc.co.uk/2/hi/health/6566953.stm.

6 Reuters Newswire Study, "Longer Needles Needed for Fatter Buttocks," November 28, 2005.

7 CBS News Sunday Morning, "Increasingly Overweight Populace Being Accommodated; Egged On?" New York, July 24, 2005.

8 N-G Gejvall, *Westerhus: Medieval Population and Church in the Light of Skeletal Remains* (Lund: Hakan Ohlssons Boktryckeri, 1960); and Pia Bennike, *Paleopathology of Danish Skeletons* (Copenhagen: Almquist and Wiksell, 1985).

9 See Greg Critser's *Fatland* (New York: Houghton Mifflin, 1993), pp. 136–39 for a fuller discussion of the different metabolism of sucrose versus fructose.

10 Nicholas Kristoff, "Dreams of Osama bin Laden," *The New York Times*, October 12, 2004.

11 John Vidal, *McLibel* (London: New Press, 1997), pp. 46–47.

12 2001 Carrot Cake on Steroids, News Release, Center for Science in the Public Interest (p.d. May 30, 2001), www.cspinet.org/new/carrotcake.html (retrieved December 15, 2008).

13 2002 Starbucks on Steroids, *870 Calories In A Drink? "Food Porn," Says CSPI*, News Release, Center for Science in the Public Interest (p.d. October 7, 2002), http://www.cspinet.org/new/200210072.html (retrieved December 15, 2008).

14 Rob Walker, "Consumed: This Joke's for You," *The New York Times*, May 4, 2008.

15 Walker, *Consumed.*

16 Carlo Colantuoni et al., "Evidence That Intermittent, Excessive Sugar Intake Causes Endogenous Opioid Dependence," *Obesity Research* 10, no. 6 (2002), pp. 478–88; and Diane Martindale, "Burgers on the Brain: Can You Really Get Addicted to Fast Food?" *New Scientist*, no. 2380, February 1, 2003.

17 J. Wang et al., "Overfeeding Rapidly Induces Leptin and Insulin Resistance," *Diabetes* 50 (2001), pp. 2786–91; and Martindale, "Burgers on the Brain."

18 A. Pocai et al., "Restoration of Hypothalamic Lipid Sensing Normalizes Energy and Glucose Homeostasis in Overfed Rats," *Journal of Clinical Investigation* 116 (2006), pp. 1081–91.

19 Anne Underwood, "How to Flunk Lunch," *Newsweek*, September 16, 2002.

20 D. T. Gilbert and J. E. J. Ebert, "Decisions and Revisions: The Affective Forecasting of Changeable Outcomes," *Journal of Personality and Social Psychology* 82 (2002), pp. 503–14; and T. D. Wilson, J. Meyers, and D. T. Gilbert, "Lessons from the Past: Do People Learn from Experience that Emotional Reactions Are Short Lived?" *Personality and Social Psychology Bulletin* 27 (2001), pp. 1648–61.

21 Ed Diener and Eunkook M. Suh, "National Differences in Subjective Well-Being," in *Well-Being: The Foundations of Hedonic Psychology*, D. Kahneman, E. Diener, and N. Schwarz, ed. (New York: Russell Sage Foundation, 1999), pp.434–50.

22 For a popular discussion of these results, see Philip Hilts, "In Forecasting Their Emotions, Most People Flunk Out," *The New York Times*, February 16, 1999. For more detail on research findings, see R. Veenhoven, "World Database of Happiness (Bibliography)" at www.world databaseofhappiness.eur.nl.

23 The *Paleo Diet*, *The Paleolithic Prescription*, and *NeanderThin* focus on re-creating what our hunter-gatherer ancestors ate. Loren Cordain, *The Paleo Diet: Lose Weight and Get Healthy by Eating the Food You Were Designed to Eat* (New York: Wiley, 2001); S. B. Eaton, M. Shostak, and M. Konner, *The Paleolithic Prescription: A Program of Diet and Exercise and a Design for Living* (New York: Harper & Row, 1988); and Ray Audette and Troy Gilchrist, *NeanderThin: Eat Like a Caveman to Achieve a Lean, Strong, Healthy Body* (New York: St. Martin's Press, 1999).

Roy L. Walford's *The 120 Year Diet: How to Double Your Vital Years* (New York: Pocket Books, 1991) [revised and republished as *Beyond the 120 Year Diet: How to Double Your Vital Years* (New York: Four Walls Eight Windows)] and a more recent book by CR Society President Brian M. Delaney and Roy Walford's daughter, Lisa Walford,

The Longevity Diet: Discover Calorie Restriction (New York: Marlowe & Co., 2005) describe using CR to extend lifespan. Any book with "macrobiotic" in the title suggests how to eat food at its freshest to maximize nutrients.

Finally, *The Perricone Prescription* has its quirky origin with a dermatologist seeking foods to delay wrinkles but ends up recommending eating lots of fish and veggies. Nicholas Perricone, *The Perricone Prescription: A Physician's 28-Day Program for Total Body and Face Rejuvenation* (New York: Collins, 2004). Even though these books begin from different premises, they all converge on similar food lists and recipes elaborating on fresh, unprocessed foods.

24 A. E. Black et al., "Measurements of Total Energy Expenditure Provide Insights into the Validity of Dietary Measurements of Energy Intake, *Journal of the American Dietetic Association* 93 (1993), pp.572–79; S. Heymsfield, D. Matthews, and S. Heshka, "Doubly Labeled Water Measures Energy Use. *Science and Medicine* 1 (1994) pp. 74–83.

25 G. Cochrane and J. Friesen, "Hypnotherapy in Weight Loss Treatment," *Journal of Consulting and Clinical Psychology* 54 (1986), pp. 489–92; J. Stradling, D. Roberts, A. Wilson, and F. Lovelock. "Controlled Trial of Hypnotherapy for Weight Loss in Patients with Obstructive Sleep Apnea," *International Journal of Obesity and Related Metabolic Disorders* 22, no. 3 (March 1998), pp. 278–81; M. Barabasz and D. Spiegel "Hypnotizability and Weight Loss in Obese Subjects," *International Journal of Eating Disorders* 8 (1989), pp. 335–41; T. Kavanagh, R. J. Shepard, and H. Doney "Hypnosis and Exercise—A Possible Combined Therapy Following Myocardial Infarction, *American Journal of Clinical Hypnosis* 16 (1974), pp. 160–65; T. Kavanagh, R. J. Shepard, V. Pandit, and H. Doney, "Exercise and Hypnotherapy in the Rehabilitation of the Coronary Patient," *Archives of Physical Medicine and Rehabilitation* 51 (1970), pp. 578–87; D. N. Bolocofsky, D. Spinler, and L. Coulthard-Morris, "Effectiveness of Hypnosis as an Adjunct to Behavioral Weight Management," *Journal of Clinical Psychology* 41 (1985), pp. 35–40; I. Kirsch, "Hypnotic Enhancement of Cognitive-Behavioral Weight Loss Treatments: Another Meta-Reanalysis," *Journal of Consulting and Clinical Psychology* 64 (1996), pp. 517–19; D. E. Thorne et al., "Are 'Fat-Girls' More Hypnotically Susceptible?" *Psychological Reports* 38 (1976), pp. 267–70.

26 David Satcher, "The Surgeon General's Call to Action to Prevent and Decrease Overweight and Obesity, Section 1.3: Economic Consequences (Rockville, MD: U.S. Department of Health & Human Services, 2001) and Section 1.2: Health Risks. See also D. B. Allison et al., "Annual Deaths Attributable to Obesity in the United States," *Journal of the American Medical Association* 282, no. 16 (October 27, 1999), pp. 1530–38 for more detail on mortality estimates.

27 Gallup Survey conducted November 3–5, 2003. News release, Gallup Organization.

28 Satcher, "The Surgeon General's Call," Section 1.3: Economic Consequences.

29 Alex Dominiguez, "Most Americans Will Be Fat Over the Long Haul: Nine Out of 10 Men and Seven Out of 10 Women Will Become Overweight," Associated Press, October 4, 2005.

30 Pew Poll Report, "Social Trends Poll: Americans See Weight Problems Everywhere but in the Mirror," (p.d. April 11, 2008) at http://www.pewtrusts.com/pdf/PRC_obesity_0406.pdf (retrieved August 8, 2006).

31 Stephanie Nebehay, "U.S. Accused of Undermining World Obesity Fight," Reuters, January 16, 2005.

32 Michael F. Jacobson, Center for Science in the Public Interest Press Conference on Free Soda in Schools, Washington DC, May 7, 1999. Transcript available at www.cspinet.org/new/free_soda2.html (retrieved May 22, 2009).

Chapter 6: Defending Home, Hearth, and Hedge Fund

1 T. H. White, *The Book of Merlyn: The Unpublished Conclusion to* The Once and Future King (Austin, TX: University of Texas Press, 1977).

2 T. H. White's *The Once and Future King* was first published as a set in 1958 in New York by Putnam, but the first three individual books had been issued earlier: *The Sword in the Stone* (New York: G. P. Putnam's sons, 1939); *The Witch in the Wood* (New York: G. P. Putnam's Sons, 1939); *The Ill-Made Knight* (New York: G. P. Putnam's Sons, 1940). *The Candle in the Wind* was published only when the set was issued in 1958.

3 R. Ardrey, *The Territorial Imperative: A Personal Inquiry into the Animal Origins of Property and Nations* (New York: Atheneum, 1966).

4 Leda Cosmides and John Tooby, *Evolutionary Psychology: A Primer* at htpp://www.psych.ucsb.edu/research/cep/primer.html (retrieved May 22, 2009).

5 For more on these patterns see Jon F. Wilkins and Frank W. Marlowe "Sex-Biased Migration in Humans: What Should We Expect from Genetic Data?" *BioEssays* 28, no. 3 (2006), pp. 290–300 (online publication date, April 1, 2006); and Isabelle Ecuyer-Dab and Michèle Robert "Spatial Ability and Home-Range Size: Examining the Relationship in Western Men and Women (*Homo sapiens*)," *Journal of Comparative Psychology* 118, no. 2 (2004), pp. 217–31 (online publication date, February 1, 2004).

6

Rank	**City/Urban area**	**Country**	**Population**	**Land area (in km²)**	**Density (people per km²)**
1	**Mumbai**	India	14,350,000	484	**29,650**
2	**Kolkata**	India	12,700,000	531	**23,900**
3	**Karachi**	Pakistan	9,800,000	518	**18,900**
4	**Lagos**	Nigeria	13,400,000	738	**18,150**
5	**Shenzhen**	China	8,000,000	466	**17,150**
6	**Seoul/ Incheon**	South Korea	17,500,000	1,049	**16,700**
7	**Taipei**	Taiwan	5,700,000	376	**15,200**
8	**Chennai**	India	5,950,000	414	**14,350**
9	**Bogota**	Colombia	7,000,000	518	**13,500**
10	**Shanghai**	China	10,000,000	746	**13,400**

From http://www.citymayors.com/statistics/largest-cities-density-125 .html (accessed October 7, 2007).

7 For White's England, the bloodiest soccer riot was not to occur until 1989 when 96 fans died in the Hillsborough disaster.

8 The trend is far from reversing; in fact, Kentucky newly passed such legislation in 2006 as did Israel in 2008. See Brandon Ortiz, "Home Intruder Law Baffles Courts. Whether It's Retroactive Is Key in Some Cases," *Herald-Leader* [Lexington, KY] at http://news.lawreader .com/?p=354 (accessed May 22, 2009) and Haaretz Service, "Knesset Approves Law Allowing Property Owners to Kill Intruders," at http:// www.haaretz.com/hasen/spages/995839.html (accessed June 25, 2008). English courts clarified existing law more firmly in that direction in

2005; see "You Can Kill a Burglar If You Have to, but not If You Want to," *The* [London] *Times* February 2, 2005.

9 The Centers for Disease Control and Prevention data published in the January 8, 2003 issue of *Journal of the American Medical Association.*

10 Department of Transportation Releases Preliminary Estimates of 2001 Highway Fatalities at http://www.nhtsa.dot.gov/portal/site/nhtsa/template.MAXIMIZE/menuitem.f2217bee37fb302f6d7c121046108a0c/?javax.portlet.tpst=1e51531b2220b0f8ea14201046108a0c_ws_MX&javax.portlet.prp_1e51531b2220b0f8ea14201046108a0c_viewID=detail_view&itemID=83d93ef851e9ff00VgnVCM1000002c567798RCRD&pressReleaseYearSelect=2002.

11 Daniel Gilbert, "If Only Gay Sex Caused Global Warming," *LA Times*, July 2, 2006. http://articles.latimes.com/2006/jul/02/opinion/op-gilbert2 (retrieved May 5, 2009).

12 Study by Sukhwinder Shergill and colleagues at the University of London, as cited in D. Gilbert, "He Who Cast the First Stone Probably Didn't," *The New York Times*, July 24, 2006, Opinion, p. 2.

13 Study by William Swann and colleagues at the University of Texas, as cited in D. Gilbert, "He Who Cast."

14 Erik Erikson, Pseudospeciation in the Nuclear Age, *Political Psychology* 6 (1985), p. 214.

15 Erik Erikson, *Identity, Youth and Crisis* (New York: W. W. Norton, 1968), p. 298.

16 Ibid.

17 Larry Beinhart, *American Hero* (New York: Random House, 1993). Republished as *Wag the Dog: A Novel* (New York: Nation Books, 2005).

18 Chris Hedges, *War Is a Force That Gives Us Meaning* (New York: Anchor Books, 2002), pp. 9, 24.

19 Hedges, *War Is a Force*, p. 5.

20 Hedges, *War Is a Force*, p. 73.

21 *Lost Rights: The Destruction of American Liberty* (New York: St. Martin's Press, 1994), p. 333.

Chapter 7: Vicarious Social Settings from Shakespeare to *Survivor*

1 Randall Stross, "Why Television Still Shines in a World of Screens," *The New York Times*, February 7, 2009, quoting the 2008 Nielsen survey.

2 Robert Kubey and Mihaly Csikszentmihalyi, "Television Addiction—How Easily We Are Harmed By Our Desires, *Scientific American*, January 25, 2002.

3 Byron Reeves and Esther Thorson's 1986 study as cited in *Television and the Quality of Life: How Viewing Shapes Everyday Experience*, ed. Robert Kubey and Mihaly Csikszentmihalyi. (Mahwah, NJ: Lawrence Erlbaum Associates, 1990).

4 T. V. Cooper et al., "An Assessment of Obese and Non-Obese Girls' Metabolic Rate During Television Viewing, Reading, and Resting," *Eating Behavior* 7 (2006), pp. 105–14.

5 Kubey and Csikszentmihalyi "Television Addiction."

6 Judith Owens et al., "Television-Viewing Habits and Sleep Disturbance in School Children," *Pediatrics* 104 (1999), electronic p. 27, htpp://www.pediatrics.aappublications.org/cgi/content/full/104/3/e27 (retrieved May 22, 2009).

7 Lillian G. Katz, "Monitoring TV Time," *Parents*, January 1989.

8 Kubey and Csikszentmihali, "Television Addiction."

9 G. K. Zipf, *Human Behavior and the Principle of Least Effort* (Reading, MA: Addison-Wesley, 1949).

10 Anemona Hartocollis, "Where You Live Can Hurt You," *The New York Times*, February 27, 2005.

11 Donna DeFalco, "A Perfect Fit: District's Physical-Education Curriculum a National Benchmark," *Naperville* (Illinois) *Sun*, March 31, 2004.

12 John Ratey and Eric Hagerman, *Spark: The Revolutionary New Science of Exercise and the Brain* (New York: Little, Brown and Company, 2008).

13 H. van Praag et al., "Running Enhances Neurogenesis, Learning, and Long-Term Potentiation in Mice," *Proceedings of the National Academy of Sciences* 96 (1999), pp. 13427–31.

14 Carl Cotman et al., *Trends in Neurosciences* 25 (2002), pp. 295–301.

15 van Praag et al., "Running Enhances Neurogenesis."

16 DeFalco, "A Perfect Fit."

17 D. Laurin et al., "Physical Activity and Risk of Cognitive Impairment and Dementia in Elderly Persons," *Archives of Neurology* 58 (2001), pp. 498–504; R. P. Friedland et al., "Patients with Alzheimer's Disease

Have Reduced Activities in Midlife Compared with Healthy Control-Group Members," *Proceedings of the National Academy of Sciences* 98 (2001), pp. 3440–45.

18 W. Stummer W. et al., "Reduced Mortality and Brain Damage after Locomotor Activity in Gerbil Forebrain Ischemia," *Stroke* 25 (1994), pp. 1862–69; and E. Carro et al., "Circulating Insulin-like Growth Factor I Mediates the Protective Effects of Physical Exercise against Brain Insults of Different Etiology and Anatomy," *Journal of Neuroscience* 21 (2001), pp. 5678–84.

19 K. R. Isaacs et al., "Exercise and the Brain: Angiogenesis in the Adult Rat Cerebellum After Vigorous Physical Activity and Motor Skill Learning," *Journal of Cerebral Blood Flow and Metabolism* 12 (1992), pp. 110–19.

20 Barry H. Cohen, "The Motor Theory of Voluntary Thinking," in *Consciousness and Self-regulation* (vol. 4), ed. R. J. Davidson, G. E. Schwartz, and D. Shapiro (New York: Plenum Press, 1986).

21 W. P. Morgan et al., "Psychological Effect of Chronic Physical Activity," *Medical Science in Sports* (1970), pp. 213–17, established that a general exercise improved depression scores in those who were clinically depressed, and J. H. Griest et al., "Running as Treatment for Depression," *Comprehensive Psychiatry* 20(1979), pp. 41–54, found the same was true for running specifically.

22 D. Scully et al., "Physical Exercise and Psychological Well Being: A Critical Review," *British Journal of Sports Medicine* 32, no. 2 (1998), p. 112.

23 James A. Blumenthal et al., "Exercise and Pharmacotherapy in the Treatment of Major Depressive Disorder," *Psychosomatic Medicine* 69 (2007), pp. 587–96.

24 Scully et al., "Physical Exercise and Psychological Well Being."

25 S. A. Paluska and T. L. Schwenk, "Physical Activity and Mental Health: Current Concepts," *Sports Medicine* 29, no. 3 (2000), pp. 167–80; E. J. Doyne et al., "Running versus Weightlifting in the Treatment of Depression," *Journal of Consulting and Clinical Psychology* 55, no. 5 (1981), pp. 748–54; and E. W. Martinsen, A. Hoffart, and O. Solberg, "Comparing Aerobic with Anaerobic Forms of Exercise in the Treatment of Clinical Depression," *Comprehensive Psychiatry* 30 (1989), pp. 324–31.

26 S. Lees and F. Booth, "Sedentary Death Syndrome," *Canadian Journal of Applied Physiology* 29, no. 4 (August 2004), pp. 447–60, and the discussion on pages 444–46.

27 News release from PRNewsWire accompanying David S. Kump and Frank W. Booth, "Sustained Rise in Triacylglycerol Synthesis and Increased Epididymal Fat Mass When Rats Cease Voluntary Wheel Running," *Journal of Physiology*, April 2005, http://jp.physoc.org/content/565/3/911.full (accessed December 15, 2008).

28 Jeff Robbins, "Technology and the Gene for Minimizing Effort," *Proceedings of the International Symposium on Technology and Society*, June 8–10, 2006, pp. 1–9.

29 Pixar's *WALL-E* to Portray our Superobese Descendants. At http://calorielab.com/news/2007/10/31/pixar-wavering-over-wall-es-portrayal-of-our-superobese-descendants (accessed May 22, 2009).

30 Stross, "Why Television Still Shines."

31 Stross, "Why Television Still Shines," quoting a December 2008 COMScore survey.

32 Jayne Gackenback, "Video Game Play and Lucid Dreams: Implications for the Development of Consciousness," *Dreaming* 16, no. 2 (June 2006), pp. 96–110.

33 James C. Rosser Jr. et al., "The Impact of Video Games on Training Surgeons in the 21st Century," *Archives of Surgery* 142, no. 2 (2007), pp. 181–86.

34 Charles Winick, "The Functions of Television: Life Without the Big Box," *Applied Social Psychology Annual* 8 (1988), pp. 217–37. Quote from p. 231.

35 Gary Steiner, *The People Look at Television: A Study of Audience Attitudes* (New York: Knopf, 1963), pp. 112–13.

Chapter 8: Intellectual Pursuits as Supernormal Stimuli

1 M. Goodman et al., "Primate Evolution at the DNA Level and a Classification of Hominoids," *Journal of Molecular Evolution* 30, no. 3 (1990), pp. 260–66.

2 Patrick D. Evans et al., "Microcephalin, a Gene Regulating Brain Size, Continues to Evolve Adaptively in Humans," *Science* 309, no. 5741 (September 9, 2005), pp. 1717–20.

3 N. Mekel-Bobrov et al., "Ongoing Adaptive Evolution of *ASPM*, a Brain Size Determinant in *Homo sapiens*," *Science* 309 (2005), pp. 1720–22.

4 B. G. Charlton, "The Rise of the Boy-Genius: Psychological Neoteny, Science and Modern Life," *Medical Hypotheses* 67 (2006), pp. 679–81.

5 B. G. Charlton, "Psychological Neoteny and Higher Education: Associations with Delayed Parenthood," *Medical Hypotheses* 69 (2007), pp. 237–40.

6 C. H. Kinsley et al., "Motherhood Improves Learning and Memory: Neural Activity in Rats Is Enhanced by Pregnancy and the Demands of Rearing Offspring," *Nature* 402 (1999), pp. 137–38; Katherine Ellison, *The Mommy Brain: How Motherhood Makes Us Smarter* (New York: Basic Books, 2005).

7 D. S. MacNutt and A. Robins, *Ximenes on the Art of the Crossword* (London: Methuen & Co Ltd, 1966), p. 49.

8 "Topics of the Times," *The New York Times*, November 17, 1924, p. 18.

9 "Condemns Cross-Word Fad," *The New York Times*, December 23, 1924, p. 17.

10 Richard Herrnstein and Charles Murray, *The Bell Curve* (New York: Free Press, 1994).

11 Rodney Brooks, in *Fast, Cheap, and Out of Control*, documentary by Errol Morris, Sony Pictures, 1997.

12 Ian Wilmut, Keith Campbell, and Colin Tudge, *The Second Creation: Dolly and the Age of Biological Control* (Cambridge, MA: Harvard University Press, 2001).

13 Andrew Pollock, "Cloning Said to Yield Human Embryos," *The New York Times*, January 18, 2008, at http://www.nytimes.com/2008/01/18/us/18embryos.html?_r=1&ex=1358398800&en=5013fc48858128da&ei=5090&partner=rssuserland&emc=rss&pagewanted=all.

14 Brian Alexander, "(You)^2," *Wired*, February 2009.

15 See the work of Mario Capecchi of the University of Utah, Sir Martin Evans of Cardiff University in Wales, and Oliver Smithies of the University of North Carolina as described at the Nobel Prize site: http://nobelprize.org/nobel_prizes/medicine/laureates/2007/press.html.

16 Jane Bosveld, "Evolution by Intelligent Design," *Discover*, March 2009, published online February 2, 2009.

17 Bosveld, "Evolution by Intelligent Design."

Conclusion: Get Off the Plaster Egg

1 See, for example, Megan Oaten's "Longitudinal Gains in Self-Control," poster presentation at The Society for Personality and Social Psychology Conference, Austin, Texas, January 2004.

2 Ed Diener and Eunkook M. Suh, "National Differences in Subjective Well-Being," in *Well-Being: The Foundations of Hedonic Psychology*, D. Kahneman, E. Diener, and N. Schwarz (New York: Russell Sage Foundation, 1999), pp. 434–50.

3 For a popular discussion of these results, see Philip Hilts, "In Forecasting Their Emotions, Most People Flunk Out," *The New York Times*, February 16, 1999. For more detail on research findings, see R. Veenhoven, "World Database of Happiness, (Bibliography)" at http://www.worlddatabaseofhappiness.eur.nl.

Illustration Credits

2 Photo of cuckoo courtesy of iStock.

4 Cartoon © The New Yorker Collection 2006 Peter C. Vey from cartoon bank.com. All Rights Reserved.

13 Photo of Niko Tinbergen painting dummy eggs by Nina Leen, TIME/ Life Collection, courtesy of Getty Images.

15 *Left*: Photo of Konrad Lorenz and imprinted goslings by Niko Tinbergen. *King Solomon's ring; new light on animal ways.* Lorenz, Konrad, New York, Crowell 1952.

15 *Right*: Drawing, self portrait with geese, by Konrad Lorenz. *Verständigung unter Tieren / (Communication among Animals)* Lorenz, Konrad, Zürich: Fontana, 1953.

32 Cartoon © The New Yorker Collection 2005 Lee Lorenz from cartoon bank.com. All Rights Reserved.

38 Two facial beauty composites from V. S. Johnston and M. Franklin, "Is Beauty in the Eye of the Beholder?" *Ethology and Sociobiology* 14, no. 3 (1993), pp. 183–99, reprinted by permission of V. S. Johnson.

47 Photo of wasp attempting to mate with orchid by Babs and Bert Wells, courtesy of The Department of Environment and Conservation, Western Australia.

53 Sketch by Konrad Lorenz, *The evolution of behavior.* Lorenz, Konrad, San Francisco: W. H. Freeman, 1958.

54 Cartoon © The New Yorker Collection 1997 Robert Weber from car toonbank.com. All Rights Reserved.

60 Cartoon © The New Yorker Collection 1994 Charles Barsotti from car toonbank.com. All Rights Reserved.

68 Photos by Adolf Naef, 1926.

71 *Left*: Sketch of older and modern teddy bears by C. Sutterlin for talk Kindersymbole: Von Puppen und "Teddies" 1976, Dresden, Germany.

71 *Right*: Anonymous German photo circa 1910.

76 Three anonymous photographic circus postcards circa 1915.

79 Photo of sign in National Park, Victoria, BC, courtesy of the Canadian Park Service.

82 Cartoon © The New Yorker Collection 1999 Jack Ziegler from cartoon bank.com. All Rights Reserved.

85 Photo of Erwin Wurm's *Fat House* and Photo of Wurm's *Fat Car*, Courtesy of the artist and Xavier Hufkens Gallery.

101 Cartoon map of United States © The New Yorker Collection 2003 Alex Gregory from cartoonbank.com. All Rights Reserved.

117 *Top*: Photos CNG coins, http://www.cngcoins.com.

117 *Bottom*: Cartoon © The New Yorker Collection 1993 Leo Cullum from cartoonbank.com. All Rights Reserved.

125 Cartoon © The New Yorker Collection 1988 Al Ross from cartoonbank .com. All Rights Reserved.

138 Scheherazade and Shahryar Illustration by Henry Justice Ford, from Lang, Andrew, ed. *Tales from the Arabian Nights*. Illustrated by H. J. Ford. 1898. G Bell & Sons London.

148 Cartoon © The New Yorker Collection 2004 Robert Mankoff from car toonbank.com. All Rights Reserved.

161 Cartoon © The New Yorker Collection 2002 Gahan Wilson from car toonbank.com. All Rights Reserved.

164 Inuit game—Anonymous engraving, circa 1910.

166 Automaton engraving in Karl Gottlieb von Windisch's *Inanimate Reason*, 1784.

174 Cartoon © The New Yorker Collection 2004 Gahan Wilson from car toonbank.com. All Rights Reserved.

Index

addiction, 17
    as model for response to pornography, 33
    to tobacco, 78
    as model for response to refined foods, 80, 86–87, 89, 91, 188*n*–89*n*
    Chris Hedges on "addiction to war," 126, 128
    television "addiction," 133, 136
    as model for storytelling and entertainment media, 137
    Anthony Cascardi on "addictive art," 137, 139
    "addictive" horror films, 143, 153
age of consent, 49–55, 185*n*
    Holland as having lowest, 49–50
    Britain, 49–50
    Canada, 49–50
    biological reasons for modern disconnection between age of sexual maturity and brain maturity, 50–51
agriculture:
    as mistake, 78
    drop in lifespan as hunter-gatherers began to farm 10,000 years ago, 80
Al Ahram, 129
Al Jezeera, 129
All in the Family, 139
Al Qaeda, 127
Alzheimer's disease, benefits of exercise and, 148, 150
anagrams, 163
animal behavior, 3, 7, 8, 24, 46, 59, 69–70
    course on, 11
    as shared interest of Tinbergen and Lorenz, 16
animé, neotenous features of, 71
anthrax:
    2001 scare, 118
    weaponized, 128
ants:
    as T. H. White's model for human aggression, 107
    as E. O. Wilson's model of social animal, 25
    horror film *Empire of the Ants,* 141–42
Arab world, 90
    McArabia sandwich and, 90
    hostility to United States after the invasion of Iraq, 90
    Moslems' Shiite vs. Sunni conflict, 118
Ardrey, Robert, *The Territorial Imperative,* 110, 191*n*
art, 169–70
ASPM gene, 161–62, 197*n*

atomic fusion, 128
automaton, 1770 chess machine, 166

Bangladesh, beans and pumpkins as desirable foods, 102
Barnes, Clive, on television, 132
barn swallow, manipulating male coloring, 3
Barnum, P. T., opinion on intellect of American consumers, 84–85
basal ganglia, and habits, 92–93
Beatles, 43
Beekvliet, 16–18
bee-wolves, 8–9
  instinct for finding nest, 9
Bell Curve, The, 171, 197*n*
Belyaev, Dmitry, 65–66
beta fighting fish, 2, 3
Bin Laden, Osama, 81, 188*n*
Bishop, Michael, 173
books, evolution of, 155–56
Booth, Frank, 150, 153, 196*n*
Bovard, James, 130
Brady Bunch, The, 139
brain:
  maturation effects of diet, 50–51
  human brain as smaller percentage of adult size at birth compared to other animals, 68
  exercise as benefiting function of, 148–49
  exercise stimulating brain-derived neurotrophic factor (BDNF), 148, 185*n*
  effects of changes in growth hormone and sex hormones on rate of growth, 160
  genetic changes and, 160–62
  microcephalin gene, 160–61
  effects of ASPM gene, 161–62
  parenthood remodeling architecture of, 162
  in relation to play and games, 162–63
  cortex for overriding simple instincts, 176–77
brain-derived neurotrophic factor (BDNF), 148
Brawndo, fictional drink "with five kinds of sugar," in film *Idiocracy,* 84
  real energy drink, 84
bridge, 165, 167
Brooks, Rod, 172, 197*n*
Brownell, Kelly, on fast food in schools, 88
Bush, George, "axis of evil" quote, 124
  Chavez insulting at UN, 124
  Chris Hedges finds ancient quotes that sound like, 127

caffeine:
  in Brawndo, 85–86
  as addictive, 86
CalorieLab, 151, 196*n*
Campbell, Keith, 172, 197*n*
cancer, 79, 80
  exercise decreasing incidence of, 150, 153
cardiovascular disease:
  high blood pressure, 78
  diet and, 78
  cholesterol levels, 78
  medical advances not impacting on death rates from, 80
  McDonald's CEO and, 83
  and saturated fat, 101
  benefits of exercise for, 147, 150, 153
careers, as powered by intellectual curiosity, 168
Cascardi, Anthony, 137, 139

Cassidy, David, as infatuation object for both men and women, 43
Cassidy, Shaun, as romantic idol, 41
Center for Science in the Public Interest (CSPI):
  monthly "food porn" award, 83–84
  survey on children's menus, 87–88
  criticism of U.S. government efforts, 98, 188*n*
charades, 165
Cheers, 139
Cheesecake Factory and CSPI "food porn" award, 83–84, 188*n*
chess, 159, 164, 167, 170
  computer programs for, 165
  the 1770 "Automaton" chess machine, 166
children:
  rape of, 34
  components of childishness as attractive in adult female faces, 37–38
  pre-pubescent crushes, 42
  percentage not offspring of their identified father, 48
  and age of consent laws, 49
  "cuteness" and, 52–59, 67–69
  and teddy bears, 70–71
  different degree of indulgence in childhood in Japan vs. United States, 73
  obesity, 76, 87
  McDonald's loses libel suit over whether it "exploited children," 82
  and fast food, 87–88, 102
  U.S. government opposition to WHO guidelines restricting advertising junk food to children, 98
  local laws against advertising junk food to children, 99
  instincts to provide resources for one's children, 114
  and war, 188
  fights, 120
  culture prolonging childhood, 123
  and television, 132–36, 157
  and sports, 146
  getting less exercise, 146–47
  parents assessing risk to, 152–53
  changes in brain after caretaking of, 162
  instinctual drive to have children, 171
  rising percentage of surviving, 171
  cloning lost children, 173
chimpanzees:
  bonobo chimps' varied sexual behavior, 48
  adopting Nigerian boy Bello, 56–57
  brain size at birth much closer to adult than for human, but less than rhesus monkey, 68
  growth hormone arriving earlier than for hominids, 160
China, aquaculture to increase rice production while farming fish on same acreage, 102
cigarettes, *see* smoking
circus side shows, 75
  fat people in, 75–76, 187*n*
cloning, 172–73, 197*n*
cocaine:
  vs. chewing coca, 78
  as name of soft drink, 86
codes, 163
computers, 157–58, 160
  Sims (Internet virtual world site), 31

computers (*continued*)
Second Life (Internet virtual world site), 31
sex in virtual reality programs, 31
televisions compared with computer use, 153
Internet, 153–55
videogames, 154, 158
Nintendo, Wii Sports and Wii Fitness, 157
Dance Dance Revolution, 158
Cosmides, Leda, 26–27, 111, 192*n*
Cosmopolitan, 35
Costa Rica, as giving up maintaining a military, 131
Couric, Katie, NBC digitally altering image of, 36
cribbage, 165
crossword puzzles, 159, 163–64
and later onset of Alzheimer's, 167
Crusades, 122
"crushes," *see* infatuations
Csikszentmihaly, Mihaly, research on television, 136
cuckoo, 1–2, 3, 171
some birds eventually learn to detect, 73
Culkin, Macaulay, as romantic idol for girls, 41
cuteness, 52–74
defining features, 52–53, 186*n*
effects on human bonding, 53–54
response across species, 54–58
"releaser" for adult nurturing, 59
Cyprus, conflict between Turks and Greeks, 118

Dance Dance Revolution, 158
Dark Shadows, Barnabas Collins, vampire character as erotic idol, 42
Dart, Raymond, 109
Darwin, Charles, 8, 182*n*
Catholic vs. Nazi view of his theories in mid-twentieth-century Austria, 16
On the Origin of Species, 24, 61
in evolutionary psychology, 24
cousin of Francis Galton, 36
unpopularity of his theories in Communist Russia in favor of Lamarckian-style inheritance of acquired traits, 64–65
motivated by intellectual curiosity, 168, 182*n*
Dawkins, Richard:
The Selfish Gene, 22, 182*n*
on Darwin, 24, 182*n*
de Beer, Sir Gavin, *Embryology and Evolution,* 67
Deliverance, 141
democracy:
inheritance of rulership even within, 114
"ensures we shall be governed no better than we deserve," 130
as "two wolves and a sheep deciding what to have for dinner," 130
diabetes, type II, 78, 79, 86
and refined sugar, 80, 101
and Pima Indians, 95
benefits of exercise for, 150
Diamond, Jared, 79, 184*n*
dice, 164
Dickens, Charles, 137
diet, 189*n,* 190*n*
prison diet, 18
and age of puberty, 50–51
effects on brain, 50–51
junk food in, 74, 77–92
diet drugs, 77
healthy, 93

"cold turkey" as easier, 90–91
changes in American last decade, 97
America vs. third world, different dietary solutions, 101–2
dietary habits:
research on changing, 92–93
brain and, 92
interactions with genes, 95
Disney, Walt:
on cuteness, 69
continuing popularity, 71
Disney World, Small World ride passengers getting larger, 75–77
dogs:
adopting children, 57
as neotenous-looking, i.e., "cute species," 59–60
as first domestic animals, 61
first fossil evidence of domestic dogs in fertile crescent, 62
in Egyptian, Assyrian, and Roman art, 63
selectively bred as pets, 63
selectively bred for work tasks, 63
as separate vs. same species as wolves, 63
in horror films *Rabid* and *Cujo,* 142
Dolly the Sheep, 172–73
domestic animals:
floppy ears and domestication, 61
bred for docility, 61–62
order in which domesticated, 61–62
dominoes, 165
Dostoevsky, Fodor, *The Idiot,* 154
"double minority" concept in defense and aggression, 118
Dracula, as erotic figure, 42–43
Duran Duran, 43
eggs:
hen laying instinctively, 2
dummy eggs as supernormal stimuli, 3, 176–77
ground nesting birds rolling, 11, 12, 13
songbird preferences for colors, markings, and size, 13
rolled out of nest as releaser of instinctive sequence, 14
domestic birds raised for, 60, 62
detection of cuckoo egg by host species, 73
desirable as small portion of human diet, 92
Eight Is Enough, 139
Elmo, neotenous features, 71
Erikson, Erik, concept of pseudospecies, 123–24, 126, 128–29
Europe, fast food spread and weight gain, 89
evolutionary psychology, 5, 26–28, 192*n*
exercise, 145–50
children getting less, 146–47
benefits for the brain and intellectual functions, 147–48
benefits for Alzheimer's, 148
as effective antidepressant, 149–50

FAA, raising estimates on average passenger weights after crashes, 77, 187*n*–88*n*
Falkland Islands, invasion of, 125
Family Circus, The, neotenous features, 71
"fat gene," 94
fear:
risk analysis, 118–19, 152
disaster films and, 141

females:
female warblers and territoriality, 11
supernormal female butterfly dummies, 15
differences in sexual instincts from men, 30–31
portrayed in male pornography, 33–35
emphasis on looks, 35
characteristics of most attractive female faces, 37
waist-to-hip ratio as an indicator of fertility, 37
women and romance novels, 41
mate choice of, 45
size of female vs. male as indicator of species mating patterns, 45–48
female fidelity vs. promiscuity across species, 48
female age of menarche dropping, 50–51, 185*n*
maternal instincts across species, 57
women's style of governing, 130
feral children adopted by animals, 56, 186*n*–87*n*
Fischer, Helen, on marriage, 45, 184*n*
food:
refining as problem, 78–80, 102
grains as low in nutrition compared to leafy plants, 80
banning junk food, 176
Food and Drug Administration (FDA) and the drink "Cocaine," 86
foxes:
Vulpes vulpes, the silver fox in Dmitry Belyaev's domestication experiments, 65–67, 68, 187*n*
France:
lines at first McDonald's, 88
French pastry, 91
Frankenstein, 143
Friends, 139
Furby, neotenous features, 71

Galton, Francis, composite photographic images, 36
geese:
eggs in research, 12, 14
Lorenz kept as pets, 14
imprinting of, 14–15
T. H. White kept as pets, 108
gender differences as leaders, 129
genetic engineering:
of food, 103
to stop war, 128
of humans, 174–75, 197*n*
ghrelin, and hunger, 87, 91
gibbons, 47–48
Gilbert, Daniel, "If Only Gay Sex Caused Global Warming" essay, 119
glucose, 80, 86, 91
storing as fat, 86
God:
Tinbergen's views on, 21
T. H. White's views on, 106
Sicilian river-god Gela, 117
as unifying force in war, 127
Godzilla, 143
Good Times, 139
gorillas, 47
Gorry, April, research on romance novels, 40, 184*n*
Gould, Stephen Jay, "A Biological Homage to Mickey Mouse," 69, 187*n*
Greenland, expedition to, 9–11
Greenpeace, and "McLibel" trial, 82–83, 188*n*
ground-nesting birds:
terns, 11
geese, 13
Guns of Navarone, The, 141

happiness research, 90, 177, 189*n*, 198*n*
Hardees' Monster Thickburger and CSPI "food porn" award, 83–84, 188*n*
Hedges, Chris, *War Is a Force That Gives Us Meaning,* 126–29, 193*n*
on war atrocities, 128
on propaganda, 129
Hello Kitty:
neotenous features, 71–72, 74
on charm bag for Shinto shrines, 73
heroin:
vs. poppy seeds, 78
withdrawal, 87, 91
addicts, 89,103
vs. ice cream, 91
Hitler, Adolf, 15–16, 130
Holland:
Dutch attitude toward hunting, 7
Nederlandse Jeugdbod voor Natuurstudie (NJN) (Dutch Youth for Nature Studies), 7, 9
homosexuality, 30, 44
and "uncle effect," 25
hormones and possible link, 48
hormones:
possible link between homosexuality and, 48
in environment skewing puberty, 50–51
sex and growth hormone pacing affecting brain development, 160, 185*n*
horror films, 141–42
hypnosis:
for weight loss, 96–97, 190*n*
differences in susceptibility, 96–97
Idiocracy and "Brawndo: The Thirst Mutilator," 84
imprinting—rigidity in geese vs. humans, 59
infatuations, 41–44
with celebrities, 41–42
1934 research on, 44
physical vs. personality traits, 44
gender differences, 44
Inuit, 9–11
their chess-like game, 164
instincts:
reflexive, 2–3
overriding with intellect, 5, 176–77
coded for a few traits, 13
Lorenz, Konrad, "On Instinct," 13
"instinct releaser," 14, 16
William James's emphasis on, 26
sexual, 30–51
for nurturing, 52–74
about territoriality, 110–13
about personal possessions, 113–14
and risk analysis, 119
to fear certain animals and events, 142
Institute of Cytology and Genetics in Novosibirsk, Siberia, 64–67
Internet, 153–55, 158, 167
and pornography, 31
letting people get around governments and media sources for information, 129
interpersonal space, 114–15
Ireland, conflict between Catholics and Protestants, 116–18
Israel:
and home protection laws, 115
soldiers shooting at rock-throwing boys, 116

Israel (*continued*)
as double minority, 118
women as possibly more able to reach solution with Palestinians, 130
ivy gourd, as source of vitamin A and B-vitamins, 103

James, William, 26, 108, 109, 123, 152
Japan, *kawaii,* and more interest in cuteness, as forerunner of world trends, 73, 98
jigsaw puzzles, 163
John, Elton, as object of infatuation, 41
Judge, Mike, 84

kawaii, Japanese term for cuteness, 72–74
Kelly, Liz, 2008 essay on infatuations and reader responses, 41–44, 184*n*
Kennedy, John F., 171
Kewpie dolls, as limit on neotenous characteristics before becoming monstrous, 69
Khan, Genghis, 170–71
Khomeini, United States as "the Great Satan," 124
"Killer Ape" hypothesis, 109–10
King Arthur, Camelot and Knights of the Round Table, 105–9
King Kong:
as erotic figure, 42
in horror films, 143
Kirby, James:
producing real-life Brawndo, 84
beverage named "Cocaine," 86
koala, as neotenous-looking—i.e., "cute" adult, 59
Kristof, Nicholas, 81, 188*n*
Krunk, Hans, biographer of Niko Tinbergen, 6, 9, 13–14, 20–22, 181*n,* 182*n*
Kubey, Robert, research on television, 136
Kuryakin, Illya, *Man from U.N.C.L.E.* character as crush object, 43
Kuwait, lines at first McDonald's, 88

Lamarckian inheritance of acquired traits, popularity in Communist Russia, 64
Large Hadron Collider, 169
Lawler, Phil, 146
Leave It to Beaver, 139
Legman, Gershorn, *Love and Death: A Study in Censorship,* 34, 183*n*
leopards:
predatory vs. nurturing instinct, 55–56
adopting humans, 58, 186*n*
leptin, and hunger, 87, 91, 189*n*
Little House on the Prairie, 139
Little League softball, 146
Lorenz, Konrad, 13–19, 23–25, 181*n,* 182*n*
personality, 14, 18
and Nazis, 15–19, 23–24
fleas' mating dance, 18–19
King Solomon's Ring, 20
Nobel Prize of, 23
views on humans, 25
On Aggression, 25
Eight Deadly Sins of Civilized Man, The, 25
eating spiders and worms, 29
on neoteny, 69
on territoriality, 111
and Sigmund Freud, 121
Lysenko, Trofim, variation on Lamarckian heritability dominating Russian biology, 64

McDonald's, 30, 81–83
"Big Mac Attack" campaign, 81
dropped after anaphylactic reaction to Big Mac, 82
"McLibel" trial, 82–83, 188*n*
sixty-year-old CEO James Cantalupo's death from a heart attack, 83
forty-three-year-old CEO Charlie Bell's death from colon cancer, 83
opening around the world, 89
McArabia sandwich, 90
Mad About You, 139
mahjong, 165
males:
barn swallows, 3
stickleback territoriality, 12
differences in sexual instincts from women, 30–31
and pornography, 33–35
wasp trying to mate with orchid, 47
testicular size as indicator of sperm competition in a species, 48
percentage of children not offspring of their identified father, 48
style of governing, 130
marriage, 45, 184*n*
romance vs. barter as determinants, 45
changes with agriculture, 45
Marx, Karl, 64, 116
masturbation, 31
percentage of men looking at pornography during, 32
synonyms for, 32
Matrix, The, 152
Mayberry, 139
Medusa, on ancient Greek coin, 117
men, *see* males
Merlyn, 105–8
Mexico, lifestyle of Pima in, 95
Mickey Mouse, increasing cuteness over time, 69, 187*n*
microcephalin gene, 160–61, 196*n*
Miller, Geoffrey, on sexual instincts and behavior, 46, 184n
monkeys:
African green ververt adopting John Ssebunya, 56–57
rhesus monkey, brain size at birth much closer to adult than for human, 68
cloning of,173
Monkichi the Monkey, on Japanese condoms, 73
morphine, 87
Morris, Desmond, *The Naked Ape,* 22, 181*n*
Mrs. Fields, cinnamon rolls and CSPI "food porn" award, 83–84, 188*n*

naloxone, and blocking opioid receptors, 87
National Geographic documentary *Eye of the Leopard,* 55–56, 186*n*
Native Americans:
flexible attitude toward territoriality, 112
Pima tribe, 94–95
nature vs. nurture debate, 21, 26–27
Nazis, 15
and drafting of Lorenz, 16
Tinbergens' resistance to in Holland, 16
N. Tinbergen's feelings about Lorenz affiliation, 19–20
imprisonment of N. Tinbergen, 16–17
neoteny, 59–74, 187*n*
definition of, 59

neoteny (*continued*)
as "releaser" for adult nurturing, 59
popular zoo animals as naturally neotenous, 59
species kept as pets bred for more neoteny, 59–60
farm animals more neotenous than wild variations, 60
biological link between neotenous behavior and docility with humans, 59–61
floppy ears and domestication, 61
corticosteroids as mediators of neoteny, 67
humans as neotenous, 67–69
as reason for intellectual pursuits, 160, 169
universities, effect on, 162
art as neotenous play, 170
news reports culled from around the world as supernormal stimulus, 153
Newton-John, Olivia, as subject of male erotic fantasies, 43
New York Times, 81, 173, 188*n*
and crossword puzzles, 163
9/11 terrorist attack, 118–19
Nippon Air:
jet painted with Pokémon characters and giving adult passengers Pikachu dolls, 72
Nobel Prize:
and Tinbergen brothers, 23
for "homologous recombination" of DNA, 174
Norworth, Jack, 144
nuclear energy, 168–69
nurturing:
instincts about, 52–54, 59, 71, 73
Pocket Pets hijacking, 74
inheritance laws and nurturing instincts, 113–14

obesity:
children and, 76, 87
world record holders for most weight, 76
equipment for, collapse-resistant toilets, double-wide ambulances, etc., 77
sugar and, 80
epidemic, 89, 97–98, 150
deaths and medical costs, 97
beneficial effects of exercise, 150
oil, ants and syrup as metaphor for wars over, 107
1001 Arabian Nights, 137
orchids, mimicking of female wasp as sexual lure, 47
"orienting response," 133–35
defined, 133
otter, as neotenous-acting—i.e., "cute" adult, 59

Palestinians:
suicide bombers, 116
as double minority, 118
women as possibly more able to reach solution with Israelis, 130
panda, as neotenous-looking—i.e., "cute" adult, 59
Parcheesi and ancient Pachisi, 164
Pavlov, Ivan, 133
P.E.4Life Institute, 146
Pets:
N. Tinbergen, attitude toward, 14
Lorenz, attitude toward, 14, 20
physical attractiveness, 36–38
facial attractiveness, 36–39
averageness as, 36, 183*n*
hormonal influence on, 37

male characteristics, 37
female characteristics, 37
proportions and, 37
BMI and, 37
cosmetics and, 38
plastic surgery, 39
as indicator of health, 39
Pikachu dolls, 72
Pima Indians, 94–95
of Arizona, 94–95
of Mexico, 95
Pizza Hut:
vending contracts in schools, 88
escaping anti-American sentiment overseas because "they think we're Italian," 89
Plato, on war, 128
play, only in young of most animals but humans and a few others do as adults, 68
Pocket Pets, as stealing nurturing resources, 74
Pokémon, neotenous features of, 71–72
poker, 165
pornography, 30–34, 182*n*–83*n*
Slade, Joseph, *Pornography in America,* 30, 183*n*
Internet porn, 31–32
and real-life rape, 33–34
children in, 34
violence in, 34
"Porn for Women," 39–40
"food porn" award, 83–84
possessions, instincts about, 113–14
power, geothermal, wind, solar, and nuclear as differentially interesting, 168–69
Precious Moments, neotenous features of, 71
prefrontal neocortex and decisions, 92–93, 96
Princeton and research on sugar-releasing opioids 86–87, 188*n*
Prozac, 149
pseudospecies:
definition of, 123
different cultures or languages as, 123
religion as, 123–24
in or out of U.S. boundaries as, 126
racial slurs as indicators of, 126
media as reinforcing or breaking down, 128–29
in war films, 141
education about, 176
psychotherapy:
for weight loss, 93–97
cognitive behavioral, 94–96
hypnotherapy for weight loss, 96–97, 190*n*
puberty, modern disconnection between age of sexual maturity and brain maturity, 50–51, 185*n*–86*n*

Radway, Janice, 41, 184*n*
rape and other sexual violence, 33–34
Ratey, John, *Spark: The Revolutionary New Science of Exercise and the Brain,* 147–48
Red Sox, fans in modern stadium seat today average 4 inches wider than those in 1912, 77, 187*n*
Researchers against Inactivity-related Disorders (RID) 150
retaliation, 120–21
rice:
as large percentage of calories consumed, 80, 99, 102–3
as fast-food base in Asia, 90
riddles, 163

risk analysis, 118
instincts and, 119
news reports and, 152
Robbins, Jeff, 150–51, 196*n*
Robin Hood, 140
robots, 172
romance novels, 35, 40–41
characteristics and behavior of hero, 40
readership and sales figures, 42
Rubik's Cube, 163
Rwandan genocide, 126

Saddam Hussein, 127
Salmon, Catherine, *Warrior Lovers,* 39, 40, 184*n*
savannahs:
as our natural environment, 3, 36
sex and romance on, 45–46, 49
plants and game there as determining our food cravings, 78
as determining what we should eat, 103
and instincts about possessions, 113–14
in relation to risk assessment, 118
as determining feared objects and animals, 142
children's natural exercise in the course of play on, 146
Scheherazade, 137–38
Schiller, Frederick, on play, 15
Scrabble, Scrabulous, and Lexulous, 166–67, 170
seal, as neotenous-acting—i.e., "cute" adult, 59
Second Life (Internet virtual world site), sexual aspects of, 30
Sedentary Death Syndrome, 150, 153, 196*n*
self-defense:
armies' structure making attacks feel like, 127
made more extreme by human identification with one another, 127
simpler for animals, 127
Sex and the City, 139
sexuality, 29–51
sexual addictions programs, 33
changes with agriculture, 48
differences in urban settings, 48
human sex with animals, 49
children and age of consent, 49–51
differing link to number of offspring in past vs. now, 171
Shaw, George Bernard, on democracy, 130
Sherlock Holmes, 137
Siberia:
as place dissidents including Darwinian biologists were sent, 64
as liberal enclave, 64
Silver, Lee, 174–75
Sims (Internet virtual world site), sexual aspects of, 30
Skywalker, Luke, *Star Wars* character as crush object, 43
Slade, Joseph, *Pornography in America,* 30, 183*n*
smoking:
nicotine withdrawal, 87
Surgeon General's warnings about, 100–101
antismoking campaign as model for antiobesity, 98, 103–4
snow buntings territoriality, 10–11
"snuff" films as myth, 34, 183*n*
soap operas, 41
sociobiology, 27–28, 116, 182*n*

Sokol, David, on proportions of house to inhabitant, 77
Solo, Han, *Star Wars* character as crush object, 43
Solo, Napoleon, *Man from U.N.C.L.E.* character as crush object, 43
Soviet Union, 64, 116
  promoting of Lamarckian-style inheritance of acquired traits over Darwin's theories, 64–65
  and expenditure of resources on military in 1970s and 1980s, 131
Spears, Britney, as subject of male erotic fantasies, 43
spectator sports, 144–45
Spurlock, Morgan, 81, 88
Starbuck's:
  White Chocolate Mocha and CSPI "food porn" award, 84
  Vanilla Crème Frappuccino coffee equal to half the calories a small woman should consume daily, 84, 188*n*
Star Wars, 137, 142
stickleback fish, 7, 19
  dummies to study territoriality, 12–13
  children's book about, 17
sugar:
  instinctive craving for 78, 97, 175, 177
  refining and concentrating, 80, 90
  diabetes and, 80
  corn syrup, 80
  five kinds in Brawndo, 84
  releasing opioids, 86–87, 188*n*
  proposed warning, 92, 98–99, 101
supernormal stimuli:
  N. Tinbergen coining term, 3
  definition of, 3–4
  as overlooked concept in evolutionary psychology and sociobiology, 27–28
  sexual, 30
  refined food as, 91
  television as supernormal social stimulus, 136
  storytelling as supernormal social stimulus, 138
  horror films as supernormal stimuli for threat, 141–42
  games as supernormal stimuli for intellectual curiosity and creativity, 159, 167
Super Size Me, documentary film, 81, 83, 88
surrender:
  animal signals for, 122
  human signals for, 122
  modern warfare masking from view, 122
Symons, Don, 179
  Warrior Lovers, 39, 40, 184*n*
  Evolution of Human Sexuality, The, 46, 183*n,* 184*n*

Tarzan, 140
teddy bears:
  evolving to be cuter across time, 69–71, 187*n*
  different preferences in infants vs. younger children, 70, 187*n*
television, 132–40, 150–52, 156–57
  average hours viewed, 132, 152
  nicknames for, 132–33
  average number in house, 133
  poll of kids on giving it up vs. giving up their fathers, 133
  violence in, 133
  attention and, 133–36
  EEG during, 134
  mood and, 135
  link to learning difficulties, 135

television (*continued*)
benefits of decreasing, 135–37, 176
as addiction, 135
as supernormal social stimulus, 136
illusion of friends with us, 139–40
Whitedot.com, 140
compared with Internet use, 153
reducing frequency of sex for couples who watch more, 156
annual "turn off your television" week, 156
"cold-turkey" quitting experiments, 156–57
territoriality, 110–13
myth that humans are territorial rather than nomadic, 110–11
Territorial Imperative, The, 110
Tinbergen's and Lorenz's views on, 110, 111
in fish and birds, 111
"home range" pattern, 112
interacting with population density, 112–13
Native Americans' flexible attitude toward, 112
testicle size, as sperm competition and the relation to female sexual patterns, 47
Tetris, 165
Thailand, ivy gourd as a desirable staple in, 103
Thatcher, Margaret, and invasion of Falkland Islands, 125
Three's Company, 139
"thrifty gene" hypotheses, 94
Tiegs, Cheryl, as subject of male erotic fantasies, 43
Tinbergen, Jan, 6, 8, 11
conscientious objector status, 16
economic sanctions, 16, 20
Niko's potential rivalry with, 21
Nobel Prize of, 23
Tinbergen, Joost, 23
Tinbergen, Lies Rutten, 9, 10, 11, 19
interaction with Lorenz, 17
Tinbergen, Luuk, 6–8, 11
achievements in ethology, 13, 22
Jewish wife and hiding from Nazis, 19
and depression, 20, 21, 22
suicide of, 22–23
Tinbergen, Niko, 3, 5, 6–25, 181*n*
parents of, 6
dissertation research, 8–9
early interest in wildlife, 7–11
writing of, 7
photography of, 7
depression of, 20, 21
as father, 20
move to Oxford University, 21
relationship with his children, 21
attitudes toward religion, 21–22
Nobel Prize of, 23
autism research and speech, 24
question of Asperger's syndrome, 24
on territoriality, 111
motivated by intellectual curiosity, 168
tobacco, *see* smoking
Tooby, John, 26–27, 111, 192*n*
transfats, 81, 92, 98
banning of, 98
Trut, Lyudmila, 66–67, 187*n*

United Nations:
helping set up home gardens in the third world, 102
Chavez's "smells of sulphur still" speaking at lectern after Bush, 124

United States:
Congress and laws about soda in schools, 9
anti-American sentiment in response to Gulf wars, 89–90
opposing world health guidelines on sugar and fat, 98
world's heaviest nation, 98
small business administration loans, 100
cartoon map of, 101
America vs. third world, different dietary solutions, 101–2
and home protection laws, 115
Constitution and Bill of Rights, 130
in or out of U.S. borders as pseudospecies distinction, 126
cannot picture disarming, 131
United States Department of Agriculture (USDA):
oversight of dietary advice, 99
grading of meats, 99–100
universities:
effect on neoteny, 162
and abstract questions, 168
University College London, research on retaliation, 120
University of Leiden:
as graduate school for Niko and Jan Tinbergen, 8
clash with Nazis over Jewish faculty, 11
L. Tinbergen thesis, 11
N. Tinbergen instructor at, 11
reopened after World War II, 19
N. Tinbergen's promotion to full professor, 21
University of Moscow:
biology department favoring Lamarckian inheritance theory, 64
Dmitry Belyaev expelled from, 65
University of Texas:
research on nuclear strike decisions, 121
publishing *The Book of Merlyn,* 191*n*

vampires:
as crush objects, 42
in horror films, 143
venereal disease, 49
videogames, 154, 158
virtual reality, sexual, 31
Vonnegut, Kurt, *God Bless You Mr. Rosewater,* 29–30, 32, 35, 182*n*

Wag the Dog:
plot resemblance to second Gulf War and George W. Bush, 124
based on first Gulf War and George Bush Sr., 125
Waistland: The R/Evolutionary Science Behind Our Weight and Fitness Crisis, 149
WALL-E, fictional portrayal of no exercise, 151–52
Walters, Barbara, CBS digitally altering image of, 36
war, 3, 74, 105–10, 116–28
World War II, 5, 102, 123
Chris Hedges sees as addictive, 126, 193*n*
in combat identification mainly with immediate fellow soldiers, 127
Trojan War, 127
World War I, 127
atrocities in, 128
whether preventable, 128
Plato on, 128
Washington Post, 2008 Liz Kelly column on infatuations and reader responses, 41–44, 184*n*

Westermarck, Edward, *The History of Human Marriage,* 45, 184*n*
White, Merry, 73
White, T. H. 105–9
  Book of Merlyn, The, 105, 191*n*
  antiwar sentiments, 105–9
  Once and Future King, The, 106
  pessimistic view of man, 106
Whitedot.com, antitelevision Web site, 140
Wild Bunch, The, 141
Wilmut, Ian, 172, 197*n*
Wilson, E. O., 25–26, 182*n*
  Sociobiology: The New Synthesis, 25
  On Human Nature, 25
wolves, adopting children, 56–59, 187*n*
  unique relationship in cooperatively hunting with man and coevolution from around 20,000 BC, 61–63
  as separate vs. same species as domestic dogs, 63
women, *see* females
World Health Organization (WHO):
  initiative on reducing fat and sugar consumption, 98
  announcement on overfeeding, 101–2
World Trade Center, attack on, 118–19
World War I, Chris Hedges quotes from, 127
World War II, effect on Tinbergen family, 5
  T. H. White opposition to, 102
  Erik Erikson's reaction to, 123
Wurm, Erin, sculptures *Fat Car* and *Fat House,* 85
Wynne, Arthur, 163

YouTube, 153

Zipf, George, *Human Behavior and the Principle of Least Effort,* 144, 194*n*
Zoloft, exercise as effective as, 149
Zorro, 140